Neuron–Glia Interrelations During Phylogeny

I. Phylogeny and Ontogeny of Glial Cells

Contemporary Neuroscience

Neuron–Glia Interrelations During Phylogeny

I. Phylogeny and Ontogeny of Glial Cells

Edited by

Antonia Vernadakis

University of Colorado Health Science Center, Denver, CO

Betty I. Roots

University of Toronto, Ontario, Canada

Humana Press ✴ Totowa, New Jersey

Printed in the United States of America. 10 9 8 7 6 5 4 3 2 1

Library of Congress Cataloging-in-Publication Data
Neuron–glia interrelations during phylogeny / edited by Antonia
   Vernadakis, Betty I. Roots.
        p. cm. — (Contemporary neuroscience)
     Includes bibliographical references and index.
     Contents: [pt.] 1. Phylogeny and ontogeny of glial cells — [pt. 2] Plasticity
     and regeneration.
     ISBN 0-89603-314-7 (pt. 1 : alk. paper)
     ISBN 0-89603-316-3 (pt. 2 : alk. paper)
        1. Neuroglia—Growth. 2. Developmental neurology.
   I. Vernadakis, Antonia, 1932–   . II. Roots, Betty I. (Betty Ida)
   III. Series.
   QP363.2.N485 1995
   591.4'8—dc20   95-15913
                                                              CIP

Preface

It is now established that neuroglia are the intimate partners of neurons and that neuronal function is a result of neuron–glia interrelations at several levels of organization. The literature shows that the study of phylogeny has contributed a deeper understanding of the complex functions of the neuroglia and the neuron–glia unit. It is the purpose of *Neuron–Glia Interrelations During Phylogeny: I. Phylogeny and Ontogeny of Glial Cells,* as well as its companion volume *Neuron–Glia Interrelations During Phylogeny: II. Plasticity and Regeneration,* to present to the scientific community a broad spectrum of information on neuroglia through phylogeny and ontogeny, the focus of this volume. In view of the role of neuroglia in plasticity and regeneration, the companion volume will cover this aspect of neuroglia during phylogeny.

Neuron–Glia Interrelations During Phylogeny: I. Phylogeny and Ontogeny of Glial Cells begins with the elegant chapter "Glial Types, Gliogenesis, and Extracellular Matrix in Mammalian CNS" by Amico Bignami, to whom this volume is dedicated. He was one of the pioneers in describing gliogenesis and this chapter brings together everything we know today on this critical topic. It also includes the latest views of Bignami on the role of extracellular matrix in gliogenesis and glial functions. "Evolution of Astrocytes in the Vertebrate CNS" by Suárez et al. complements and extends the information in Bignami's chapter by including ependymal astrocytes. The role and fate of radial glia is well documented in several chapters and in fact the chapter by Voigt and de Lima is dedicated to these cells. The following few chapters begin the saga of astrocytes up the phylogenetic tree. Several interesting views are put forward in these articles. More importantly, they lay out the similarities and differences in neuroglial lineages in phylogeny. "Astrocyte Differentiation and Correlated Neuronal Changes in the Opossum Superior Colliculus" by Cavalcante and Barradas discusses astrocytes and the heterogeneity of radial glia in marsupials and "Neuron/Glia Lineages During Early Nervous System Development in Amphibian and Chicken Embryos" by Cochard et al. considers in depth both neuronal and glial lineages in amphibian and avian CNS. A provocative view in this chapter is the description of the role of the notochord in the expression of oligodendrocytes. "The Neuroglia in the CNS of Teleosts" by Lara et al. describes in depth the ependymocytes, which are the principal neuroglial type found in lower vertebrates (teleosts). Lara et al. conclude that neuroglia in the teleost CNS "possess some characteristics that permit their inclusion in some of the traditional neuroglia types together with particularities that could correspond to evolutive and/or functional adaptations of the teleosts." Neurogenesis and gliogenesis in *Drosophila* are well presented in "Glial Interactions with Neurons During *Drosophila* Development" by Fredieu and Mahowald. This chapter presents evidence that glial cells are important in the CNS of the fly.

Astrocyte development in higher mammals, including humans, is presented in the chapter by Elder. This article attempts to clarify the dilemma among scientists using lower mammals (rat) as to whether rules applying to the rat glial development also apply to the higher mammalian systems. A special feature of the present volume is "Schwann Cells in Phylogeny," by Stewart and Jessen, which documents the evolution of Schwann cells and their invertebrate counterparts and discusses how Schwann cell-axon relationships and functions have changed during phylogeny. The final two chapters of the volume are dedicated to the evolution of myelinogenesis. "The Evolution of Myelinating Cells" by Betty I. Roots takes the reader through the phylogenetic tree from the annelida and arthropoda to mammals. The final chapter, by Jeserich, Stratmann, and Strelau, describes in depth myelinogenesis in the trout.

In conclusion, *Neuron–Glia Interrelations During Phylogeny: I. Phylogeny and Ontogeny of Glial Cells* presents a comprehensive treatise of the evolution of neuroglia including Schwann cells and neuron–glia interrelations in lower as well as in higher animal phyla. This volume has encyclopedic value not only for neurobiologists, but all scientists interested in the evolution of the nervous system.

Antonia Vernadakis
Betty I. Roots

Dedication

"I am a part of all that I have met.
And this grey spirit yearning in desire
To follow knowledge like a sinking star,
Beyond the utmost bound of human thought."
—Alfred, Lord Tenneyson

Amico Bignami was an outstanding scientist and scholar. One of his remarkable qualities was a great respect for those "who were here first," such as Virchow, Weigert, Ramon y Cajal, del Rio Hortega. Nearly every lecture, seminar, or comment would be prefaced by such "an old observation." This is the sign of a true scholar and places Bignami among the field's pioneers.

He had an amazing grasp of the complexity of cell interactions, and a talent for presenting the information in a simple language. Although the discovery of GFAP as a reliable marker for astrocytes was important—indeed the technique has become routine for neurobiologists—his contributions to our understanding of astrocytes are far more comprehensive, ranging from the genesis of glial cells to their differentiation, to their normal functions, to their role in disease and regeneration. In each of these steps, his views have not always been universally accepted, but he always argued, effectively, that his scheme of the cellular events from the genesis of precursors to the production of a differentiated astrocytes cannot be disproved. He has many followers.

Bignami was an endless source of information, primarily because he knew brain embryology and development. He was one of those rare species of scientists who appreciated cell organization and cell structure. He was frequently called upon to express his opinion about some fresh report in the literature concerning a new cell marker or to correlate an in vitro finding with some possible in vivo counterpart. The phrase "Bignami agrees" was a stamp of approval.

Amico Bignami (1930–1994) was born in Montreux, Switzerland. He received his medical training at the University of Rome, Italy and was Associate Professor of Pathology there from 1959 to 1969. He was Associate Professor of Pathology at Stanford University from 1969 to 1976. He had been Professor of Neuropathology at the Harvard Medical School since 1976 and was Staff Neuropathologist and Director of Spinal Cord Injury Research as well as Associate Chief of Staff for Research and Development at the Brockton/West Roxbury Veterans Administration Medical Center.

He had great vision and was always seeking new horizons, the latest being his contribution to understanding the role of extracellular matrix in gliogenesis and glial differentiation, as well as in regeneration. Amico Bignami left a legacy in the field of neuroscience. His creative mind will be greatly missed. The scientific community has suffered a great loss, although we are richer for his contributions.

Antonia Vernadakis
Betty I. Roots

Contents

Astrocyte Differentiation and Correlated Neuronal Changes in the Opossum Superior Colliculus
Leny A. Cavalcante and Penha C. Barradas

Neuron/Glia Lineages During Early Nervous System Development in Amphibian and Chicken Embryos
Philippe Cochard, Cathy Soula, Marie-Claude Giess, Françoise Trousse, Françoise Foulquier, and Anne-Marie Duprat

Glial Interactions with Neurons During *Drosophila* Development
John R. Fredieu and Anthony Mahowald

PART II: PHYLOGENY OF MYELINATION
The Evolution of Myelinating Cells
Betty I. Roots

A Cellular and Molecular Approach to Myelinogenesis in the CNS of Trout
Gunnar Jeserich, Astrid Stratmann, and Jens Strelau

Contents for the Companion Volume

Neuron–Glia Interrelations During Phylogeny:
II. Plasticity and Regeneration

xiii

Contributors

JOSÉ AIJÓN • *Departamento de Biología Celular y Patología, Universidad de Salamanca, Salamanca, Spain*

JOSÉ R. ALONSO • *Departamento de Biología Celular y Patología, Universidad de Salamanca, Salamanca, Spain*

PENHA C. BARRADAS • *Instituto de Biologia, UERJ, Rio de Janeiro, Brazil*

*AMICO BIGNAMI • *Harvard Medical School and Brockton/West Roxbury, Department of Veterans Affairs Medical Center, Boston, MA*

GUILLERMO BODEGA • *Departamento de Biologia Celular y Genética, Universidad de Alcalá, Madrid, Spain*

LENY A. CAVALCANTE • *Instituto de Biofísica Carlos Chagas Filho, Universidade Federal de Rio de Janeiro, Rio de Janeiro, Brazil*

PHILIPPE COCHARD • *Centre de Biologie du Développement, Université Paul Sabatier, Toulouse Cedex, France*

ANA D. DE LIMA • *Max-Planck-Institut für Entwicklungsbiologie Spemannstr, 35/I, Tübingen, Germany*

ANNE-MARIE DUPRAT • *Centre de Biologie du Développement, Université Paul Sabatier, Toulouse Cedex, France*

GREGORY A. ELDER • *Brookdale Center for Molecular Biology, Mt. Sinai Medical Center, New York, NY*

BENJAMIN FERNÁNDEZ • *Departamento de Biologia Celular, Universidad Complutense, Madrid, Spain*

FRANÇOISE FOULQUIER • *Centre de Biologie du Développement, Université Paul Sabatier, Toulouse Cedex, France*

JOHN R. FREDIEU • *Department of Cell Biology and Anatomy, Oregon Health Sciences University, Portland, OR*

MARIE-CLAUDE GIESS • *Centre de Biologie du Développement, Université Paul Sabatier, Toulouse Cedex, France*

GUNNAR JESERICH • *Department of Animal Physiology, University of Osnabrück, Osnabrück, Germany*

*Deceased.

xv

KRISTJAN R. JESSEN • *Department of Anatomy and Developmental Biology, University College London, London, United Kingdom*

JUAN M. LARA • *Departamento de Biología Celular y Patología, Universidad de Salamanca, Salamanca, Spain*

ANTHONY MAHOWALD • *Department of Molecular Genetics and Cell Biology, University of Chicago, Chicago, IL*

BETTY I. ROOTS • *Department of Zoology, University of Toronto, Toronto, Ontario, Canada*

MIGUEL RUBIO • *Departamento de Biologia Celular y Genética, Universidad de Alcalá, Madrid, Spain*

CATHY SOULA • *Centre de Biologie du Développement, Université Paul Sabatier, Toulouse Cedex, France*

HELEN J. S. STEWART • *Department of Anatomy and Developmental Biology, University College London, London, United Kingdom*

ASTRID STRATMANN • *Department of Animal Physiology, University of Osnabrück, Osnabrück, Germany*

JENS STRELAU • *Department of Animal Physiology, University of Osnabrück, Osnabrück, Germany*

ISABEL SUÁREZ • *Departamento de Biologia Celular y Genética, Universidad de Alcalá, Madrid, Spain*

FRANÇOISE TROUSSE • *Centre de Biologie du Développement, Université Paul Sabatier, Toulouse Cedex, France*

ALMUDENA VELASCO • *Departamento de Biología Celular y Patología, Universidad de Salamanca, Salamanca, Spain*

THOMAS VOIGT • *Max-Planck-Institut für Entwicklungsbiologie Spemannstr, 35/I, Tübingen, Germany*

PART I

PHYLOGENY AND ONTOGENY OF GLIAL CELLS

Glial Types, Gliogenesis, and Extracellular Matrix in Mammalian CNS

Amico Bignami

1. Introduction

In brain and spinal cord, neurons and their processes (axons and dendrites) are almost completely ensheathed by glia with one important exception, i.e., the synaptic junctions mediating the chemical transmission of nerve impulses (Figs. 1 and 2). The central nervous system (CNS) is completely surrounded by glia not only on the surface, but also at the interface between blood vessels and the nervous tissue (Figs. 3 and 4). It is only at the site of entry of spinal and cranial nerves that the glial membrane and the overlying basal lamina are perforated (Fig. 5). In view of this anatomical arrangement, two general concepts have arisen concerning the role of glia in the CNS: Neuroglia, or, more specifically, astrocytes are the equivalent of fibroblasts in peripheral organs and provide the brain with a connective tissue; and, neurons and glia form a single tissue and this also in view of the fact that they have a common origin in the primitive neuroepithelium. From this perspective neurons and glia function as a unit.

There are two major types of neuroglia, astrocytes and oligodendrocytes, the myelin-forming cells. Resident macrophages as opposed to blood-borne macrophages, are the third type of glia (microglia). Astrocytes and oligodendrocytes (like neurons) originate from the neuroepithelium. It is generally believed that microglia derive from blood-borne macrophages invading the brain early in postnatal development.

From: *Neuron-Glia Interrelations During Phylogeny: I. Phylogeny and Ontogeny of Glial Cells*
A. Vernadakis and B. Roots, Eds. Humana Press Inc., Totowa, NJ

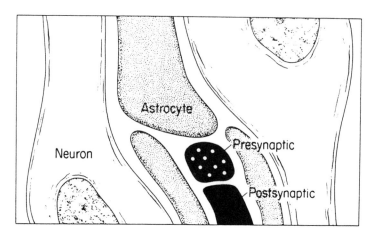

Fig. 1. In gray matter, astrocytes surround neurons and their processes. The extracellular space between astrocytes and neurons and between astrocytes and myelin (*see* Fig. 2) is filled by an amorphous extracellular matrix, i.e., a matrix that does not contain structural elements (e.g., collagen fibrils) as in other parts of the body (*see* Fig. 13). This matrix is soluble in the aqueous fixatives used for electron microscopy leading to the morphological appearance of an empty extracellular space extremely reduced in size (5% of the brain by volume vs 17–20% according to physiological measurements). Neurons are in direct contact with each other at the synapse. The synaptic cleft is filled with neuronal surface proteins, the neuronexins (Ushkaryov et al., 1992).

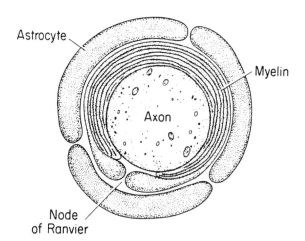

Fig. 2. In white matter, astrocytes mainly surround myelin sheaths, an oligodendrocyte product. It is only at the nodes of Ranvier that astrocytes "see" the periaxonal space.

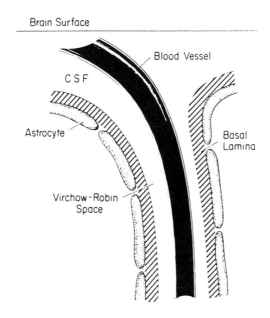

Fig. 3. Astrocyte processes form a continuous lining on the surface of the brain and of blood vessels entering the brain from the leptomeninges. A basal lamina is always interposed between the astrocyte lining and mesenchyma (blood vessels and leptomeninges). A perivascular space communicating with the leptomeninges and containing cerebrospinal fluid (the Virchow-Robin space) is the equivalent of a lymphatic space in the brain. It is the site of inflammatory cell accumulation in meningo-encephalitis. As shown in Fig. 4, the perivascular space does not extend to capillaries.

Astrocytes were discovered by Virchow (1821–1902), a German pathologist who dissected the periventricular layer of the human brain and observed that it was made of a tissue different from nerve but distinctive to the brain (Weigert, 1895). Virchow later came to the conclusion that the tissue was not confined to the periventricular layer but extended without boundaries into the brain itself, filling the interstices between neurons and separating neurons from blood vessels. He called this tissue neuroglia (neural glue) because he believed that it served the purpose of binding neurons together. It should be noted that both astrocytes and oligodendrocytes are presently included under the name of neuroglia.

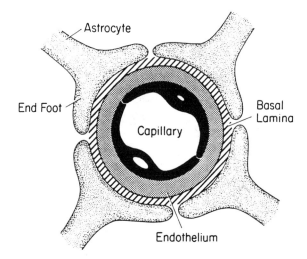

Fig. 4. Astrocyte processes (endfeet) abutting on a brain capillary and forming the glial perivascular membrane. Note the basal lamina between endothelial cells and astrocyte endfeet. In mammalian vertebrates the blood-brain barrier is owing to the presence of *tight* junctions between endothelial cells. Putative functions of the astrocyte endfeet are to provide substances that induce tightening of the blood-brain barrier during development, transport to the circulation of potassium released during neuronal activity, and production of hemostatic substances.

Virchow also discovered one of the main properties of astrocytes, i.e., the capacity of forming scars in the brain and spinal cord. After hardening the tissue with chromic acid, the white matter of spinal cord dorsal columns in a case of tabes had lost under the microscope, the granular appearance of myelin and a mesh of densely packed extremely fine fibrils had become apparent. Tabes is a form of neurosyphilis primarily affecting the dorsal roots at their entry into the spinal cord. Myelinated axons in the dorsal columns degenerate as a result and astrocytes react with a process called isomorphic gliosis because they orient their fibers in the same caudo-cranial direction as the sensory axons undergoing Wallerian degeneration.

Considering the techniques available to Virchow one may understand Deiters' statement that the discovery of neuroglia appeared to be the result of divination rather than based on demonstrable facts (Weigert, 1895). Incidentally, it was Deiters who

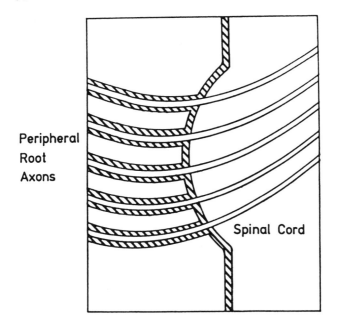

Fig. 5. Real openings in the basal lamina (interrupted line) and under-lying glial membrane surrounding the CNS only occur where axons enter or leave the CNS. This zone is called the nerve root and can be divided into a peripheral part and a CNS part. In the peripheral part, axons are ensheathed by Schwann cells, the peripheral glia. Schwann cells not only form myelin but also the basal lamina surrounding myelin. On entering the spinal cord, axons lose their basal lamina. Myelin is now formed by oligodendrocytes. (There are major differences in com-position between the myelin formed by Schwann cells and oligoden-drocytes.) Astrocytes surround the myelin sheaths in the CNS. They do not form a basal lamina (*see* Fig. 2), but secrete an extracellular matrix filling the space between astrocytes and myelin (*see* Fig. 13A).

In anterior motor roots, the axons originate inside the spinal cord, whereas in posterior (sensory) roots they originate in the dorsal root ganglia outside the spinal cord. If sensory axons are crushed in the periph-eral root they will behave as in peripheral nerve, i.e., they will regener-ate vigorously, but they fail to enter into the spinal cord (Stensaas et al., 1979; Bignami et al., 1984; Stensaas et al., 1987).

in 1865 first described the microscopic appearance of glial cells in the CNS.

Astrocytes were identified in tissue sections by Cajal's gold chloride sublimate, staining the whole cell body and its processes, and by Weigert's method, staining the astrocytic processes that contain glial fibrils. The small glial cells that did not stain with the gold chloride sublimate method for astrocytes were called "third element" by Cajal. Del Rio Hortega developed a silver carbonate impregnation method that stained the processes of these small cells, and mainly on morphological criteria he was able to show that they comprised two kinds of cells: oligodendroglia, the myelin-forming cells and microglia, the resident CNS macrophages (Penfield, 1932).

1.1. Glial Cell Markers

The traditional methods of conventional histology have been largely superseded by immunocytochemical methods. We will briefly review the markers most commonly used for the identification of the different types of glial cells in tissue sections and cell culture.

As a general rule, intracellular antigens are preferable to cell surface antigens for studies conducted on tissue sections. Among cytoplasmic antigens intermediate filament (IF) proteins are easier to use because the antigenicity usually withstands fixation. Furthermore, they are relatively insoluble so that diffusion of the reaction product is not a problem, as in the case of S-100 protein. Cell surface markers are particularly useful for the study of nonfixed (nonpermeabilized) cells in monolayer culture.

1.1.1. Astrocytes

1.1.1.1. GLIAL FIBRILLARY ACIDIC PROTEIN

Glial fibrillary acidic protein (GFAP) is most commonly used as a marker for the identification of astrocytes in vivo and in vitro (Bignami et al., 1972; Dahl and Bignami, 1985). For practical purposes, astrocytes may be defined as GFAP-positive cells.

Astrocytes, like neurons, are characterized by the presence in their cytoplasm of fibrils that are formed at the electron microscopic level by 10-nm filaments. The other two types of glial cells, oligodendrocytes and microglia, do not contain these filaments.

Two major neurohistological stains developed at the turn of the century, i.e., Weigert's stain for astrocytes and Cajal's silver nitrate neurofibrillary method, are based on the selective decoration of glial filaments and neurofilaments respectively, and it was first suggested by Weigert (1895) that the selectivity of the stain was owing to the specificity of the substance forming the fibrils. In more recent times, antibodies to the proteins forming the filaments have become essential tools for the identification of neurons and glia (Dahl and Bignami, 1985).

The cytoskeleton in eukaryotic cells comprises three types of filamentous structures: microtubules, approx 25 nm in diameter; actin filaments, 5–7 nm in diameter; and 10-nm filaments, also referred to as IFs, because their width lies between that of microtubules and microfilaments. Microtubules and microfilaments play an important role in a variety of cell functions, such as mitosis, motility, phagocytosis, and secretion. The function of the intermediate filaments is still basically unknown although indirect evidence suggests that they may play a role in stabilizing the cell shape.

If IFs probably are not essential to cell function, their constitutive proteins have become very important for the identification of the cell type in vivo and in vitro. Actin and tubulins, the subunits of microfilaments and microtubules, respectively, are evolutionarily conserved proteins and similar if not identical in different cell types. Conversely, intermediate filaments in different tissues, although morphologically similar, are made of distinct proteins based on immunochemical criteria. Intermediate filament proteins may thus be considered as taxonomic characters allowing the identification of the cell type independently of morphological criteria, which is particularly important in development, in tissue culture, and in tumors.

The following IF proteins are currently used as immunocytochemical markers for the identification of the cell type: the neurofilament triplet (neurons), GFAP (astrocytes), vimentin (mesenchyma), desmin (muscle), and keratins (epithelia).

In apparent contrast with the immunological findings, amino acid sequence analysis has shown marked similarities allowing a general structural model of IF proteins. The proteolysis-resistant α-helical middle domain forming the IF backbone is the most conserved part of the molecule, whereas the more variable nonhelical aminoterminal and carboxyterminal domains probably account for the different properties of IFs in different cell types. Surprisingly, most astrocyte-specific epitopes of GFAP reside in the proteolysis-resistant α-helical domain (Dahl et al., 1986).

GFAP is not strictly confined to astrocytes and related cells in the CNS. GFAP immunoreactivity has been reported in enteric glial (Jessen and Mirsky, 1980), Schwann cells of unmyelinated peripheral axons (Yen and Fields, 1981; Dahl et al., 1982), lens epithelium (Hatfield et al., 1984), and Kupffer cells in the liver (Gard et al., 1985).

As evidenced by monoclonal antibody studies, there appear to be subtle differences between "central" and "peripheral" GFAP. For example, a monoclonal antibody only stained a subpopulation of enteric glia outside the CNS (Jessen et al., 1984). Since "central" and "peripheral" GFAP appear to be coded by the same gene (Mokuno et al., 1989), the immunological differences between the two proteins are probably caused by posttranslational modifications.

1.1.1.2. S-100 PROTEIN

Comparison of starch gel electrophoresis patterns of concentrated fractions from DEAE-cellulose chromatography of liver and brain extracts showed that the brain contained small, highly acidic, water soluble proteins that were absent in liver extracts and that also appeared brain specific by immunological criteria. One of these proteins was named S-100 because of its solubility in saturated ammonium sulfate (Moore, 1969). S-100 is part of a group of dimeric calcium-binding proteins that includes calmodulin, troponin-C, and parvalbumin. All these proteins exhibit a common structural motif called the EF-band where E and F stand for two α-helices flanking a loop of 12 amino acids (14 in the case of S-100 protein) that binds calcium. Recent evidence suggests that S-100 protein plays a role in the Ca^+-dependent regulation of the astrocyte cytoskeleton (Bianchi et al., 1993).

In the CNS, S-100 protein is mainly restricted to astrocytes in accordance with an early report of its persistence in thalamic nuclei undergoing retrograde neuronal degeneration (Cicero et al., 1970). Careful comparative studies on the distribution of S-100 protein and GFAP in adult murine brain have shown a similar distribution of the two antigens (Ludwin et al., 1976). Compared to GFAP antibodies, astrocytic perikarya are better visualized with S-100 antisera, whereas astrocytic processes appear shorter (Ghandour et al., 1981; Björklund et al., 1983).

1.1.1.3. GLUTAMINE SYNTHETASE

The astrocytic localization of glutamine synthetase (GS) an enzyme that catalyzes the amidation of glutamate to glutamine, was first reported by Norenberg (1979). The distribution of GS immunoreactivity in rat brain closely corresponds to that observed with anti-GFAP except for the presence of GS in the ependyma, the epithelial-like cells lining the surface of the cerebral ventricles. Astrocytes responding to traumatic injury accumulate both GFAP and GS but the amount of GS immunoreactive material gradually decreased 3 wk after injury. Old reactive astrocytes packed with GFAP filaments were almost completely GS negative (Norenberg, 1983). GS immunoreactivity thus may be considered a good indicator of astrocyte function.

The cortisol induction of GS has been reported in mouse primary astrocytic cultures (Juurlink et al., 1981) and in differentiating Müller glia of the avian embryonic neural retina (Linser and Moscona, 1979, 1983; Norenberg et al., 1980).

1.1.2. Oligodendrocytes

Compared to astrocyte markers, and particularly GFAP, oligodendrocyte markers are not extensively used for in vivo studies. This is probably because of the fact that the major product of oligodendrocytes (myelin) is readily recognized on morphological criteria and that compared to reactive astrocytes reactive oligodendrocytes are not a prominent feature in brain injury.

1.1.2.1. GALACTOCEREBROSIDE

Galactocerebroside (GC), a major myelin lipid, is the most commonly used marker for the identification of oligodendrocytes in vitro (Raff et al., 1978). The usefulness of GC as an oligodendrocyte marker depends on the fact that it appears relatively early in oligodendrocyte differentiation but at a stage when the cell is irreversibly committed to the oligodendrocyte lineage. Expression of myelin proteins, i.e., myelin-associated glycoprotein, myelin basic protein, and myelin proteolipid is a later phenomenon (Arenander and de Vellis, 1994).

1.1.2.2. O ANTIGENS

Monoclonal antibodies reacting with a series of cell surface molecules designated O antigens are used on studies of oligodendrocyte development in vitro because they recognize progressively more differentiated stages (Trotter and Schachner, 1989). Monoclonal antibody O1 identifies cells committed to the oligodendrocyte lineage, whereas O4-positive cells are still capable of becoming astrocytes in the presence of fetal calf serum. The O4 antigen is the most commonly used antigen of the series. It recognizes sulfatide, seminolipid, and an unidentified antigen (Bansal et al., 1989).

1.1.2.3. CARBONIC ANHYDRASE

First identified by Giacobini (1961) in the CNS, the glial form of carbonic anhydrase (isoenzyme C) is predominantly localized in oligodendrocytes and in Müller glia of the retina (Kumpulaien et al., 1983). Carbonic anhydrase B, the other main enzyme has been demonstrated only in erythrocytes and vascular walls.

1.1.2.4. GLYCEROL PHOSPHATE DEHYDROGENASE

Glycerol phosphate dehydrogenase (GPDH) is another enzyme predominantly localized in oligodendrocytes (Arenander and de Vellis, 1994). Like glutamine synthetase, GPDH is cortisol inducible in both tissue culture and in the intact organism. Glycerol phosphate dehydrogenase has been also identified in Bergmann glia of cerebellum. There appear to be species differences in this respect since Bergmann glia was better stained by GPDH antibodies in mice than in rat (Fisher et al., 1981).

Interestingly, GPDH expression in Bergmann glia depends on the interaction with the adjoining Purkinje cells. Glycerol phosphate dehydrogenase immunoreactivity disappeared concomitant with the loss of Purkinje cells in several mouse mutants (Fisher, 1984).

1.1.3. Microglia

Although brain resident macrophages may have unique properties, there is still no stain available that is specific for microglia. For the identification of microglia, lectin and antibodies that recognize surface markers of macrophages elsewhere in the body are used (Table 1). Several of these markers are only expressed when microglia respond to injury. Furthermore, reactive microglia like macrophages can be identified by their ability to engulf particles (Ling and Wong, 1993) and by the presence on their cell surface of receptors for acetylated low-density lipoproteins (ac-LDL). These can be visualized with a fluorescent marker bound to ac-LDL (Guilian et al., 1989).

Table 1
Surface Leukocyte Antigens
Used for Immunostaining of Reactive Microglia

Major histocompatibility complex (MHC) class I and class II glycoproteins (McGeer et al., 1993). Rarely detected in resting microglia.
Leukocyte common antigen (LCA) or CD45 (Sedgwick et al., 1991).
Variable expression in resting microglia. LCA has been recently
identified as a tyrosine phosphatase and thus belongs to a family of
transmembrane proteins called tyrosine phosphatase receptors,
although the nature of the ligand(s) remains unknown
(Trowbridge, 1991).
Complement type 3 receptor (Perry et al., 1993). Stains both resting
and reactive microglia.
Receptor to the FC chain of immunoglobulins, particularly the FcγR1
receptor (McGeer et al., 1993). Stains both resting and reactive
microglia.

1.1.4. Radial Glia

The mesenchymal IF protein *vimentin* is perhaps the best
marker to identify immature glia, particularly radial glia in tissue
sections (Dahl and Bignami, 1985). Vimentin is coexpressed
with GFAP by astrocytes in vitro so that vimentin-positive glial
cells in tissue culture are only considered immature in the absence
of GFAP.

Immediate filament associated proteins are proteins that
copurify with IF proteins and colocalize with IFs by immunostaining but are unable to assemble into IFs. One of these proteins
(IFAT-70/280 kDa) has been used successfully to stain radial glia
in tissue sections (Yang et al., 1993). A major advantage compared
to vimentin is that the IF-associated protein disappears when
radial glia differentiate into GFAP-positive cells. Conversely,
vimentin is carried over to some extent in mature white matter
(but not gray matter) astrocytes.

Two antibodies that have been extensively used by
Caviness and his collaborators in morphological studies of
radial glia are monoclonals *RC1* and *RC2* (Edwards et al., 1990;
Takahashi et al., 1990). The antibodies were derived from mice
immunized with homogenates of rat embryo brain relatively late
in development (d 15–17). The RC antigens have not been completely characterized.

External Surface

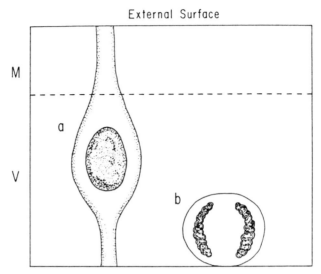

Ventricular Surface

Fig. 6. The neural tube. **(a)** Cylindrical bipolar cell with the nucleus deep in the wall of the neural tube synthesizing DNA. This cell is the precursor of all cells forming the CNS, with the possible exception of microglia. **(b)** Rounded cell in mitosis near the ventricular surface. The daughter cells will assume the cylindrical shape and start again the generation cycle. M, Marginal zone from which cell nuclei are excluded during their interkinetic movements towards and away from the ventricular surface.

Antibodies to cell surface gangliosides have been also used to identify immature glia in vitro. These markers will be discussed in the next section (gliogenesis).

1.2. Gliogenesis

1.2.1. The Neural Tube (Fig. 6)

Neurons and glia derive from the neural tube. All cells of the neural tube look alike under the electron microscope. His (1889) believed that the early neural tube was formed of precursors of

1. Glial cells (spongioblasts) stretching across the full thickness of its wall;
2. Round germinal cells seen in mitosis close to the ventricular surface; and
3. Neuroblasts.

According to Schaper (1897), the early neural tube was made of a single cell type (pseudostratified columnar epithelium) viewed either in interphase or mitosis. Although His' views were generally accepted at the time, Schaper's interpretation was essentially confirmed by kinetic analyses, e.g., DNA content of the nuclei, autoradiography after injection of [3H] thymidine (reviewed by Sidman and Rakic, 1973). As shown in Fig. 6, a columnar epithelial cell synthesizing DNA preparatory to division withdraws its externally directed cytoplasmic process, rounds up at the ventricular surface, and divides. Each postmitotic cell reextends a cytoplasmic process toward the external surface. The bipotential neuroepithelial cells do not stain for vimentin but express a protein called nestin (Lendhahl et al., 1990) a new member of the IF family.

1.2.2. Radial Glia (Fig. 7)

The identification of neuroglia in the wall of the neural tube is difficult on morphological grounds because embryonal glia (like neuroepithelium) have an elongated shape spanning the entire thickness of the neural tube, and for this reason they are called radial glia. Before the introduction of IF proteins as cell markers, the identification of radial glia depended on the Golgi method and electron microscopy. The most distinctive feature is the presence of endfeet abutting on the external surface and on newly formed blood vessels. It is possible that by their first interaction with mesenchymal tissue primitive neuroepithelial cells lose the ability of generating neurons and thus transform into radial glia.

Radial glia stain well with antibodies to vimentin, the mesenchymal IF protein, with antibodies to an IF-associated protein (IFAP-70/280) and with antibodies to still incompletely characterized antigens (RC1, RC12), probably sulfatides. As to the expression of GFAP in radial glia, there appear to be discrepancies in the literature. Some of these discrepancies may be explained by the fact that there seem to be species differences as to the time of expression of GFAP. As an example, GFAP is expressed earlier in human (Antanitus et al., 1976) and monkey (Levitt and Rakic, 1980) CNS than in murine brain (Bignami and Dahl, 1974). It should be noted however, that these differences may be interpreted with caution since it is often difficult to determine the equivalent developmental stage in species that differ significantly in gestation

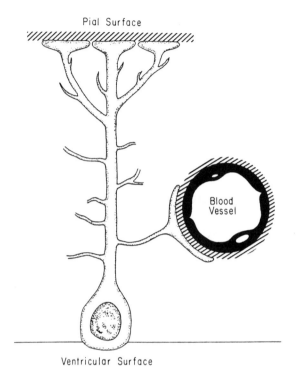

Pial Surface

Blood Vessel

Ventricular Surface

Fig. 7. The drawing shows a radial glial cell spanning the entire thickness of the neural tube. The cell would be hardly distinguishable from a primitive neuroepithelial cell if it was not for the endfeet abutting on the surface and on blood vessels at a later stage of development. Another difference is that radial glia (but not primitive neuroepithelium) express vimentin, the mesenchyma-type IF protein. Primitive neuroepithelial cells are vimentin-negative, but express nestin another IF protein. Furthermore, radial glia-like mature astrocytes in gray matter contain abundant particulate glycogen. Glycogen can thus be used to identify radial glia if the brain is fixed with solutions that allow the preservation of glycogen in tissue sections (Gadisseux and Evrard, 1985). Radial glial cells are the main type of glia in lower vertebrates. In mammalian brain glial cells of radial appearance persist in the retina (Müller glia, *see* Fig. 12A) in the hypothalamus (tanycytes) and in the cerebellar cortex (Bergmann glia).

time and degree of maturity at birth. Another reason for the discrepancy is that there is a direct correlation between the time of appearance of any given antigen and the sensitivity of the immunohistochemical technique used to detect the antigen, i.e., the antigen appears earlier with more sensitive techniques. Very sensitive techniques, however, may be misleading, not only because of the risk of artifact but also because structural antigens produced in minute amounts may not be used by the cell.

The role of radial glia for the guidance of migrating neurons first suggested by Cajal was convincingly demonstrated by the elegant electron microscopic work of Rakic and his collaborators showing the close apposition of migrating neuroblasts to radial glia in cerebral cortex (reviewed by Sidman and Rakic, 1973). Although in the cerebral hemispheres, the cells guiding neuronal migration are an embryonal form of glia, migration of granule neurons in postnatal cerebellum occurs along radial glial fibers that persist in adulthood (Bergmann glia) and appear to be mature at the time of neuronal migration, as evidenced by morphological appearance and GFAP immunohistochemistry (Bignami and Dahl, 1973).

It should be noted, however, that not all neuronal migrations occur along radial glial fibers. As a specific example, neuroblasts forming the germinal layer of the cerebellum migrate from the caudolateral margin of the fourth ventricle to the surface of the cerebellum in close contact with preexisting axonal bundles (Hynes et al., 1986). This suggests that the surface requirements for migrating neurons are not stringent, or, in other words, that migrating neuroblasts can use different surfaces as long as they lead in the right direction. Furthermore, extensive neuronal migration parallel to the ventricular surface and thus crossing at right angles the radial glia is suggested by the generation of widespread cerebral cortical clones (Walsh and Cepko, 1992). In this study, rats were injected with retrovirus carrying distinctive DNA inserts at the time of neuroblast division and the genetic tags were identified by the polymerase chain reaction (PCR) at the end of neuronal migration.

Finally, little is known concerning the interactions between radial glia and migrating neurons. The adhesive proteins that appear to be important for axonal fasciculation do not seem to be involved. The only antibodies that blocked neuron-glia binding were those against astrotactin, a neuronal antigen missing in *weaver* mice (Hatten, 1990). Weaver is a mutation characterized by defective neuronal migration in the cerebellum. Astrotactin has not been sequenced and thus we do not know where it stands compared to other surface proteins.

1.2.2.1. THE FATE OF RADIAL GLIA

Two lines of evidence suggest that radial glia differentiate into astrocytes.

1. Although it is often difficult to construct cell lineages on morphological grounds, the transitional forms between radial glia and astrocytes observed by Schmechel and Rakic (1979) in monkey telencephalon during development, are extremely suggestive for a transformation of radial glia into mature astrocytes.
2. In human and monkey, GFAP can be used as a marker to identify radial glia (Antanitus et al., 1976; Levitt and Rakic, 1980). Furthermore, like protoplasmic astrocytes in rat cerebral cortex, radial glia can be identified with GFAP antibodies when they respond to injury (Bignami and Dahl, 1974).

If astrocytes originate from radial glia, the question may be asked whether they are the same astrocytes that appear in white matter during "myelination gliosis." No answer is available to this question, nor to the more general question whether there are different types of astrocytes. For the purpose of experimentation the possibility may be considered that there are two types of astrocytes: astrocytes originating from radial glia in the gray matter, and astrocytes originating from astroblasts that divide locally at the time of myelination gliosis and populate white matter.

1.2.3. Myelination Gliosis

As originally reported by Roback and Scherer (1935), myelination is preceded by a burst of cell division, the so-called "myelination gliosis" (Fig. 8). Most axons in nonmyelinated white matter lie close together without intervening glia (Fig. 9). Conversely, neurons and their processes are enveloped by glia in the mature CNS except at the synapses. Myelination is a late phenomenon mainly occurring postnatally in mammalian development.

It should be emphasized that most glial cells in the CNS originate by local division. Conversely, most neuroblasts divide in germinal layers and then migrate to their final destination. A major question is whether a precursor of the two major types of glia (astrocytes and oligodendrocytes) exists in the CNS, as suggested by in vitro studies, or whether astrocytes and oligodendrocytes originate from different precursors. In myelination gliosis, GFAP-posi-

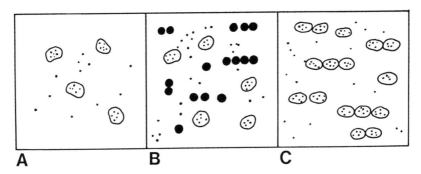

Fig. 8. The drawing shows cerebral white matter stained with a routine method (hematoxylin-eosin) before **(A)**, after **(B)**, and during **(C)** myelination. With this method, only the nuclei of glial cells are recognizable. Glial cell processes are undistinguishable from the other components of white matter (myelin and axons). Only few glial nuclei are seen in nonmyelinated white matter (A). In mature white matter (B) two types of nuclei are present: densely stained nuclei (oligodendrocytes) and lighter stained nuclei (astrocytes). If the plane of the section is parallel to an axonal tract (e.g., pyramidal tract, corpus callosum) oligodendrocyte nuclei are often aligned in short rows, as shown in the drawing. It should be noted that identification of the glial type on the basis of nuclear characteristics is not foolproof e.g., cells that appear to be astrocytes may turn out to be oligodendrocytes by electron microscopy. The most characteristic feature of white matter in "myelination gliosis" is a greatly increased nuclear density (C). In sections conducted parallel to an axonal tract, the nuclei are arranged in characteristic long rows. "Myelination gliosis" starts before the appearance of GFAP and myelin, the major products of astrocyte and oligodendrocyte differentiation. Compared to oligodendrocytes in mature brain, the nuclei are larger and surrounded by basophilic material, indicating the presence of nucleic acids and active protein synthesis.

tive cells (astrocytes) and myelin-forming cells (oligodendrocytes) appear in short succession (Bignami and Dahl, 1973), thus suggesting the existence of a common precursor in accordance with the in vitro findings. However, the evidence presently available indicates that there is no such common precursor. The two major lines of evidence are as follows.

1. Lineage studies using replication defective retroviruses that carry a β-galactosidase gene (Sanes et al., 1986; Cepko, 1988) have shown that most clones stained for the galactosidase

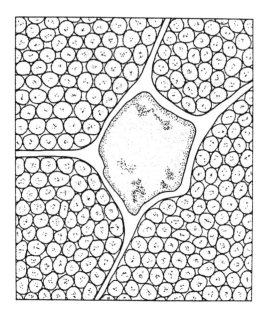

Fig. 9. The drawing shows the transverse section of an axonal tract before myelination. A glial cell (in the center of the drawing) contacts a few axons but for the most part axons contact each other (after myelination, the axons will be surrounded by myelin sheaths and astrocytes, *see* Fig. 2). Interactions between adhesive proteins on the surface of nonmyelinated axons are probably important for fasciculation, the process by which axons serving similar functions bundle together during development forming fiber tracts (e.g., the pyramidal tract). The surface proteins involved in the phenomenon (L1, NCAM) belong to the immunoglobulin class of adhesive molecules, i.e., they share sequence homology with immunoglobulins. The adhesive interactions are either homophilic e.g., NCAM-NCAM, or heterophilic, e.g., NCAM-L1 (Linnemann and Bock, 1989). Nonmyelinated fiber tracts are among the few regions that do not contain hyaluronic acid in immature brain (Bignami and Asher, 1992). Because of its hydrophilic properties, hyaluronic acid occupies a considerable amount of space and as such, it would prevent direct interactions between adhesive molecules on the cell surface.

gene product are formed by either astrocytes or oligodendrocytes (Luskin et al., 1988; Price and Thurlow, 1988; Vaysse and Goldman, 1990). It should be noted that these studies are conducted in late fetal or early postnatal murine brain where most of the cells capable of division are glial cells.

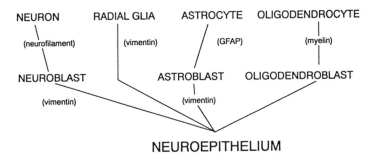

Fig. 10. A simplified view of cell lineage in the CNS. The terms neuroblast, astroblast, and oligodendroblast indicate cells committed to the neuronal, astrocyte, and oligodendrocyte lineage, respectively, but still capable of dividing. In the case of neurons, division in germinal layers is followed by migration. Glial cell precursors continue to divide after migration. Intermediate filament proteins (neurofilament, vimentin, GFAP) and myelin lipid (galactocerebroside) are the main taxonomic markers. Regardless of the cell lineage, most immature cells express vimentin, neuroepithelium being the main exception.

(Hippocampus and cerebellum are exceptional in that the neuronal precursors of the small granule cells continue to divide postnatally in most mammalian species.) If the brain is infected earlier in development, clones formed of both glia and neurons occur, as one would expect (Fig. 10).

2. Autoradiography studies, after injection of [³H] thymidine in rat optic nerve, indicate that during "myelinating gliosis" most astrocytes are generated before the oligodendrocytes (Skoff, 1990). If astrocytes and oligodendrocytes were derived from a common precursor, one would expect the two cell types to appear simultaneously.

The next question to be raised is the origin of the relatively few precursors of astrocytes and oligodendrocytes that reside in nonmyelinated white matter and divide to form the adult population of astrocytes and oligodendrocytes. A peculiar feature of the nonmyelinated neonatal brain in rat as well as in human is that it contains a large population of small cells in the subependymal zone of the lateral ventricles, the so-called germinal layer (Fig. 11). Since neurogenesis has ceased in both species except in cerebellum and hippocampus, it is tempting to speculate that

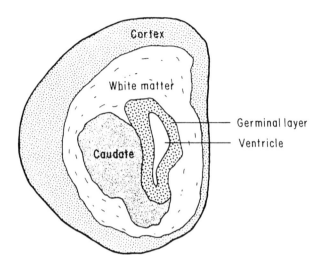

Fig. 11. Transverse section through the frontal lobe of a newborn rat. The frontal horn of the lateral ventricle is surrounded by a thick layer of small vimentin-positive cells called the germinal layer. Since migration of neurons to the cerebral cortex and to the caudate nucleus has ceased at this stage of development, it is reasonable to assume that the germinal layer is formed by immature glial cells. It is generally believed that the astrocyte and oligodendrocyte precursors undergoing mitosis in nonmyelinated white matter derive from the germinal layer. The cells that do not migrate probably differentiate into astrocytes, as evidenced by the presence of a subependymal GFAP-positive glial layer in mature brain.

migration from the subependymal zone into nonmyelinated white matter accounts for the presence of astrocyte and oligodendrocyte precursors in this location. In newborn spinal cord, however, small round cells are not found around the central canal (the equivalent of the ventricle) and it is difficult to believe that periventricular cells migrate all the way from the lateral ventricle down to the spinal cord. Since myelination occurs at an earlier stage in the spinal cord, it is still possible that migration occurs at an earlier stage and that no immature cells are left in the periventricular zone at the time of birth. Furthermore, it must again be emphasized that in view of the remarkable degree of *local* mitotic activity during myelination gliosis, glial migration in development is certainly a modest phenomenon compared to neuronal migration, which only occurs when neurons have left the mitotic cycle.

1.2.4. Gliogenesis in Optic Nerve and Retina

Gliogenesis in the optic nerve and retina is an interesting phenomenon also because it sheds some light on gliogenesis elsewhere in the brain. The retina derives from an evagination of the primitive neuroepithelium, which, like in brain and spinal cord, differentiates into neurons and radial glia. Different from brain and spinal cord, radial glia, which in the retina is called Müller glia, maintains its radial appearance throughout adult life (Fig. 12A).

There is another type of glia in the retina, astrocytes in the layer of optic nerve fibers, which form the optic nerve. Lineage studies suggest that these astrocytes do not originate from the retinal epithelium. Clones derived from the retina are formed by neurons and radial glia and apparently they do not contain astrocytes (Turner and Cepko, 1987). It is thus generally believed that astrocytes in the nerve fiber layer of the retina come from the optic nerve. Astrocytes probably enter the retina together with blood vessels. Cultures of the retina do not yield astrocytes before vascularization (Watanabe and Raff, 1988) and astrocytes are confined to vascularized regions of the retina (Schnitzer, 1988).

In the optic nerve, like in other white matter tracts, most astrocytes and oligodendrocytes derive from local mitoses at the time of "myelination gliosis" (Skoff, 1990). At this time a ventricular cavity is no more present in the evagination of the brain vesicles, which forms the retina and the optic nerve. It is thus believed that the mitotic precursors of the oligodendrocytes and astrocytes migrate into the optic nerve from the subependymal germinal layer in the brain. If this is true, there are species differences in the extent of migration of the oligodendrocyte precursors as judged by the extent of myelination. In some species (e.g., human, rat) myelination stops at the lamina cribrosa, a sieve in the sclera through which bundles of optic nerve axons leave the eye (Fig. 12B). In other species (e.g., rabbit, dog) myelination extends beyond the lamina cribrosa to reach the optic nerve head.

1.2.5. Gliogenesis In Vitro

If the lineage and autoradiographic data indicating the existence of different glial precursors are confirmed this could indicate that there are major differences between gliogenesis in vivo and in vitro since a common astrocyte-oligodendrocyte precursor has been convincingly demonstrated in tissue culture (Raff et

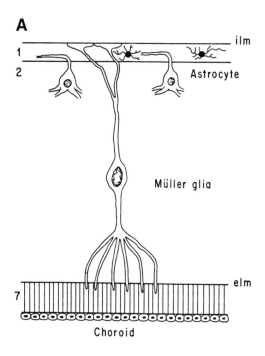

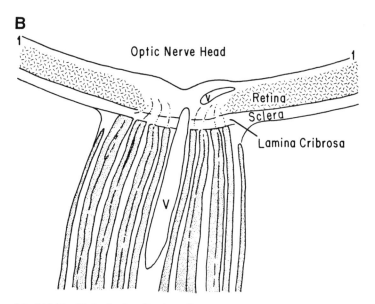

Fig. 12. **(A)** Radial glial cells (Müller glia) are the main type of glia in the retina. They extend from the vitreous surface where their expanded bases form the internal limiting membrane (ilm), to the external limit-

al., 1983). Primary monolayer cultures derived from neonatal rat optic nerve develop into monolayer of fibroblast-like vimentin-positive cells, most of which stain with GFAP antibodies (type 1 astrocytes). Growing on top of these cultures there is a population of small, biphase dark cells. These small round cells will differentiate into oligodendrocytes and stellate astrocytes (type 2 astrocytes) in serum-free and serum-containing tissue culture media, respectively. Type 2 astrocytes are characterized by the presence on their surface of a ganglioside recognized by monoclonal antibody A2B5. The A2B5 antigen, the main marker presently available for type 2 astrocytes, is also expressed by neurons, and this is one of the reasons why the original work on gliogenesis using this antibody has been done on optic nerve cultures. Optic nerves do not contain neurons. Unfortunately, the A2B5 antibody does not work reliably in tissue sections and it is thus difficult to tell whether type 2 astrocytes also exist in vivo (Miller et al., 1989).

Enriched cultures of the oligodendrocyte-type 2 astrocyte precursor can be obtained by the procedure of McCarthy and de Vellis (1980), a method routinely employed by investigators working on oligodendrocytes in vitro. By this procedure, the small round cells growing on top of monolayers derived from primary dissociated cultures of perinatal murine brain are shaken off, allowed to adhere to plastic, and subcultured, the main contaminant of the preparation being microglia. Purified microglia can be obtained from these cultures since they adhere more strongly to plastic than the oligodendrocyte precursors. These can be removed by gently shaking the flasks, whereas more strongly

---

ing membrane (elm) at the base of the rod and cone layer (layer 7). Only short Müller glia processes extend beyond the elm around the bases of rods and cones. Stellate astrocytes are only found in the layer of optic nerve fibers (1) originating from large neurons in the ganglion cell layer (2). **(B)** Astrocytes do not originate in the retina but migrate from the optic nerve together with blood vessels (V) when the retina becomes vascularized. In several species optic nerve fibers are *not* myelinated beyond the lamina cribrosa, a series of sieve-like membranes arranged transversely across the scleral canal, thus suggesting that oligodendrocyte precursors are unable to migrate beyond this point. 1, Layer of optic nerve fibers.

adhering microglia are released by vigorous shaking (Giulian and Baker, 1986).

Perhaps the most interesting results obtained in tissue culture studies of the glial precursor is that in serum-free cultures, platelet-derived growth factor (PDGF) stimulates cell division and delays differentiation of the precursors into mature oligodendrocytes (Noble et al., 1988; Richardson et al., 1988). Without PDGF, the precursors will still differentiate into oligodendrocytes but more rapidly and without expansion of the cell population. This observation may lead to the understanding of myelination gliosis, one of the major developmental landmarks in postnatal brain maturation. Considering that astrocyte maturation precedes oligodendrocyte maturation, it is tempting to speculate that astrocytes provide the growth factors that are required for proliferation of the oligodendrocyte precursor and myelination gliosis (Raff et al., 1988). Another possibility is that PDGF secreted by nonmyelinated axons stimulates proliferation of the oligodendrocyte precursor. The hypothesis is based on the finding that PDGF is mainly produced by neurons and that the predominant cells expressing the PDGF receptor are oligodendrocyte precursors (Yeh et al., 1993). It thus remains to be seen whether myelination gliosis is initiated by astrocytes or neurons.

1.3. Formation of Extracellular Matrix

An extracellular matrix containing acidic glycosaminoglycans is a prominent feature of the invertebrate nervous system. The highly charged interstitium is believed to play a role in the distribution of water and electrolytes between the intravascular and interstitial compartments (Abbott and Lane, 1986).

Until recently little was known concerning the nature of the extracellular matrix in the vertebrate CNS. The reason why so little research was done on the subject, especially in view of the explosion of extracellular matrix (ECM) research outside the brain, is probably to be found in the ultrastructural appearance of CNS tissue. In electron micrographs of conventionally fixed and embedded material, brain extracellular space appears as a thin cleft accounting for less than 5% of the brain volume. In fact the morphological evidence was so compelling that to use the words of J. G. Nicholls in a Citation Classics published in *Current Contents* in 1989, it led to "fanciful ideas" such that "the intracellular fluid of glial cells represented the extracellular space surrounding neurons." It should be noted, however, that there was no other way to reconcile the morphological with the experimental evidence, such as the large chloride space demonstrated by Vernadakis and Woodbury (1965). Another puzzling phenomenon was that the space appeared empty as if its content had been extracted during tissue processing.

Following the seminal paper of van Harreveld et al. (1965), it is now generally accepted that the small size of brain extracellular space in electron micrographs of conventionally processed tissue is an artifact probably owing to cellular swelling resulting from asphyxia, the real extracellular space in mature brain being approx 17–20% in volume (Nicholson and Rice, 1986).

Cell processes and extracellular space are well beyond the power of resolution of the light microscope, and it is thus remarkable that Cajal (1897) and Golgi (1903) were able to identify a perineuronal extracellular matrix. It was called "capa de cemento" or "ciment pericellulaire" by Cajal and later became known as Golgi's pericellular net. It soon became apparent that the Golgi's net was not confined to the surrounding of neurons, but also extended without interruption throughout the CNS including white matter and that it closely corresponded to the glial reticulum described by Held. In fact, Held (1902) recognized two substances in the glial net: a pale homogeneous substance that probably corresponds to the glial cytoplasm, and a granular more intensely stained substance that he identified as Golgi net. More recently, the glial origin of the Golgi perineuronal net was demonstrated by Lafarga et al. (1984) in the fastigial nucleus of the rat cerebellum.

It is not possible in this brief historical introduction to review the extensive work conducted on brain ECM with conventional histochemical methods. Suffice it to say that the presence of hyaluronidase-sensitive mucopolysaccharides (glycosaminoglycans) in brain interstitium was shown in several studies (Palladini and Alfei, 1968; Deeler and Nakanishi, 1983). The impact of this work however was negatively affected not only by the electron microscopic appearance of CNS tissue, but also by the belief that the cationic dyes used in these studies such as Alcian blue (Scott, 1985) lacked specificity (Ripellino et al., 1985).

1.3.1. Hyaluronic Acid

Hyaluronic acid (HA) plays the main structural role in the formation of brain extracellular matrix as the backbone of protein aggregates. This explains why the extracellular space appears empty by electron microscopy. Hyaluronic acid is readily dissolved during the preparation of tissues for electron microscopy. How this matrix is anchored to the cell surfaces lining brain extracellular space still remains to be determined.

Hyaluronic acid, the only glycosaminoglycan not covalently bound to protein, is synthesized on the inner surface of the cell membrane and then extruded to the outside of the cell during elongation in both eukaryotic cells and bacteria (Prehm, 1986). The HA synthetase gene from group-A Streptococcus-pyogenes recently has been cloned and sequenced (DeAngelis et al., 1993). The other glycosaminoglycans (chondroitin sulfate, keratan sulfate, heparan sulfate) are made in the Golgi vesicles that are not present in bacteria and then covalently linked to protein (Carney and Muir, 1988).

1.3.2. HA-Binding Proteins

The two HA-binding proteins isolated from brain in milligram quantities are a 60 kDa glycoprotein called glial hyaluronate-binding protein (GHAP) because its localization at the light-microscopic level is similar to that of GFAP, and a 365-kDa chondroitin sulfate proteoglycan (Perides et al., 1989, 1992). Both proteins are closely related to versican, a fibroblast proteoglycan preferentially expressed in precartilaginous mesenchyma during development (Zimmermann and Ruoslahti, 1989; Shinomura et al., 1992; Bignami et al., 1993a,b). In fact, amino acid sequence analysis indicates that GHAP corresponds to the amino-terminal HA-binding region of versican (Zimmermann and Ruoslahti, 1989). Experiments aimed at generating GHAP from versican are at present under way in this laboratory. Recently, we obtained promising results with stromelysin, a matrix metalloproteinase secreted in an inactive form requiring other proteinases for activation in vivo (Matrisian, 1990).

The evidence presently available suggests that versican is not the only aggregating proteoglycan in the brain. The antigen reacting with monoclonal antibody Cat-301 (McKay and Hockfield,

1982) has been recently identified and found to be similar to aggrecan, the HA-binding proteoglycan of cartilage extracellular matrix (Fryer et al., 1992).

Compared to versican in peripheral tissues, brain versican appears considerably smaller because of the lesser number and/or length of the chondroitin sulfate side chains. It is well established that variations in the number and length of glycosaminoglycan side chains are a major source of diversity among proteoglycans possessing the same core protein (Hardingham and Fosang, 1992). The finding of a lesser chondroitin sulfate content in brain versican is consistent with studies on the diffusion of ions in brain interstitium that do not suggest the existence of a highly charged ECM (Nicholson and Rice, 1986).

1.3.3. CD44

Figure 13 illustrates *hypothetical* models of the extracellular matrix in white matter and gray matter. According to these models, hyaluronic acid (HA) is anchored by a receptor to the cell surface of astrocytes lining the extracellular space in both white matter and gray matter.

Hyaluronate binding sites were first identified on the surface of many cells by Underhill and Toole (1979) and later characterized as an 85-kDa membrane protein (Underhill et al., 1985). Recently, it was found that the hyaluronate receptor is identical to CD44, a cluster of differentiation (CD) antigens that have been implicated in the homing of lymphocytes to mucosal lymphoid tissues (for review, *see* Underhill, 1992).

Among its diverse functions, the HA receptor plays an important role in retaining HA on the cell surface with formation of the pericellular coats that are characteristic of several cell types in culture (Toole, 1991). In chondrocytes, this pericellular coat serves as a scaffold for the retention of HA-binding proteins secreted by the cell (Knudson, 1993). Evidence available in this laboratory suggests that astrocytes in vitro are similar to chondrocytes in this respect in that they produce HA, express CD44 and are capable of retaining HA and HA-binding proteins on their surface (Asher and Bignami, 1991, 1992). It should also be noted that CD44 (the HA receptor) is also expressed by astrocytes in vivo as evidenced by immunohistochemical studies (Girgrah et al., 1991; Vogel et al., 1992).

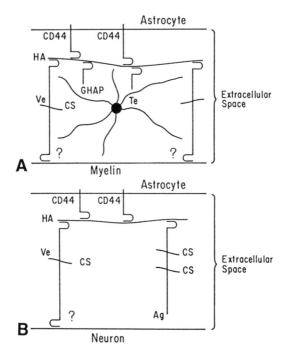

Fig. 13. Hypothetical model of the matrix filling the extracellular space between astrocytes and myelin in white matter **(A)** and between astrocytes and neurons in gray matter **(B)**. In both white matter and gray matter, astrocytes secrete hyaluronate (HA) which is retained on their surface by a hyaluronate receptor (CD44). Hyaluronate in turn binds versican, a chondroitin sulfate proteoglycan. In white matter hyaluronate also binds (GHAP), a proteolytic product of versican corresponding to its hyaluronate-binding region. At its opposite end versican binds myelin in white matter and neurons in gray matter, as evidenced by the fact that, unlike GHAP, versican is not completely released from the tissues by hyaluronidase digestion. Tenascin is another ECM protein that has been identified in white matter. The six-armed tenascin (hexabrachion) binds to versican and probably "reinforces" the hyaluronate-protein aggregates, as suggested by the fact that the myotendinous junction, an area of great mechanical stress, is one of its main locations. Aggrecan, the hyaluronate-binding proteoglycan of mature cartilage, participates in the formation of the pericellular nets that surround certain types of CNS neurons. In the brain, aggrecan contains more chondroitin sulfate proteoglycans than versican, thus suggesting that it may provide neurons with a negatively charged interstitium that may retain calcium in their immediate surroundings.

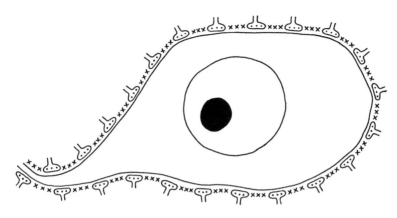

Fig. 14. Golgi's perineuronal nets are formed by the concentration of extracellular matrix around neurons. The net-like appearance is owing to the exclusion of the matrix from the synaptic clefts. Compared to the surrounding interstitium, perineuronal extracellular matrix is more negatively charged, because of its higher content of chondroitin sulfate (*see* Fig. 13B). Such fixed negative charges may interfere with the movement of cations, the effect being more pronounced with divalent calcium. Interestingly, parvalbumin-positive neurons in the cerebral cortex are often surrounded by Golgi's nets (Celio, 1993). Parvalbumin is a calcium-binding protein that acts as a calcium sink in rapidly relaxing skeletal muscle (modified from Celio and Blümcke, 1994).

As shown in Fig. 13, there are some differences between white matter and gray matter with respect to the proteins that form the aggregates. Glial hyaluronate-binding protein is mainly a white matter protein, thus suggesting that the enzyme responsible for the cleavage of versican is not activated in gray matter. Aggrecan is strictly confined to gray matter where it participates in the formation of Golgi's perineuronal nets (Fig. 14). In white matter the aggregates are "reinforced" by tenascin (Perides et al., 1993), a protein that is typically found at the myotendinous junctions and that probably binds versican.

In Fig. 13, versican is anchored to the neuronal and myelin surface with a question mark. The experimental observation supporting this hypothesis is that versican differs from GHAP and aggrecan in that it is not completely removed from the tissue by hyaluronidase digestion. In view of its similarity at the carboxy-terminal end with adhesive proteins (Zimmermann and Ruoslahti,

1989), versican may function as the adhesive connection between extracellular HA and cell surfaces.

1.3.4. Function

As to the function of brain ECM, we expect that as in other tissues (Laurent and Fraser, 1992) hyaluronic acid in the brain regulates water flow and contributes to water homeostasis. The HA compartment will exclude macromolecules but not low molecular weight solutes. In fact, a function for HA as a space filler in the brain is consistent with the notion that brain ECM does not interfere with the diffusion of ions, including ions with a high charge density such as calcium (Nicholson and Rice, 1986). Most polar groups in hyaluronic acid are involved in intramolecular interactions, thus preventing intermolecular interactions (Scott et al., 1984). As to the possibility that HA prevents the diffusion of macromolecules in brain interstitium, the experiment reported by Bairati in 1953 is interesting. When bovine spinal cord was injected with India ink, diffusion within the tissue was only observed in the presence of hyaluronidase.

Compared to peripheral versican, brain versican contains fewer/shorter glycosaminoglycan side chains and thus fewer negative charges, as evidenced by the small change in molecular weight occurring after chondroitinase digestion. The perineuronal Golgi's net (Fig. 14), however, comprises an aggrecan-like proteoglycan that is more negatively charged than brain versican (but still much less charged than cartilage aggrecan) judging from its chondroitin sulfate content (unpublished observations). Interestingly most cortical neurons surrounded by a Golgi net are positive for parvalbumin, a calcium-binding protein (Lüth et al., 1992). A possibility to be considered is that the negatively charged perineuronal net is another mechanism to interfere with the diffusion of calcium.

References

Abbott JN Lane NJ (1986) The blood-brain inferface in invertebrates. Ann NY Acad Sci 481:20–45.

Antanitus DS Choi BH Lapham LW (1976) The demonstration of glial fibrillary acidic protein in the cerebrum of the human fetus by indirect immunofluorescence. Brain Res 103:613–616.

Arenander AT de Vellis J (1994) Development of the nervous system, in Basic Neurochemistry: Molecular, Cellular and Medical Aspects. 5th ed. (Siegel GJ et al., eds.), Raven, New York, pp. 573–605.

Asher R Bignami A (1991) Localization of hyaluronate in primary glial cultures derived from newborn rat brain. Exp Cell Res 195:401–411.

Asher R Bignami A (1992) Hyaluronate binding and CD44 expression in human glioblastoma cells and astrocytes. Exp Cell Res 203:80–90.

Bairati A (1953) Spreading factor and mucopolysaccharides in the central nervous system of vertebrates. Experientia 9:461–462.

Bansal R Warrington A Gard AL Ransch T Pfeiffer SE (1989) Multiple and novel specificities of monoclonal antibodies O1, O4 and R-mAB used in the analysis of oligodendrocyte development. J Neurosci Res 24:548–557.

Bianchi R Giambanco I Donato R (1993) S-100, but not calmodulin, binds to the glial fibrillary acidic protein and inhibits its polymerization in a Ca^{2+} dependent manner. J Biol Chem 268:12669–12674.

Bignami A Asher R (1992). Some observations on the localization of hyaluronic acid in adult newborn and embryonal rat brain. Int J Dev Neurosci 10:45–58.

Bignami A Dahl D (1973) Differentiation of astrocytes in the cerebellar cortex and the pyramidal tracts of the newborn rat. An immunofluorescence study with antibodies to a protein-specific to astrocytes. Brain Res 49:393–402.

Bignami A Dahl D (1974) Astrocyte-specific protein and radial glia in the cerebral cortex of newborn rat. Nature 252:55–56.

Bignami A Chi NH Dahl D (1984) Regenerating dorsal roots and the nerve entry zone. Immunofluorescence study with neurofilament and laminin antisera. Exp Neurol 85:426–436.

Bignami A Hosley M Dahl D (1993a) Hyaluronic acid and hyaluronic acid-binding proteins in brain extracellular matrix. Anat Embryol 188: 419–433.

Bignami A Perides G Rahemtulla F (1993b) Versican, a hyaluronate-binding proteoglycan of embryonal precartilaginous mesenchyma is mainly expressed postnatally in rat brain. J Neurosci Res 34:97–106.

Bignami A Eng LF Dahl D Uyeda CT (1972) Localization of the glial fibrillary acidic protein in astrocytes by immunofluorescence. Brain Res 43: 429–435.

Björklund H Dahl D Haglid K Rosengren L Olson L (1983) Astrocytic development in fetal parietal cortex grafted to cerebral and cerebellar cortex of immature rats. Dev Brain Res 9:171–180.

Cajal S Ramón (1897) La red superficial de las células nerviosas centrales. Rev Trimestral Micrografica 3:199–204.

Carney SL Muir H (1988) The structure and function of cartilage proteoglycans. Physiol Rev 68:858–910.

Celio MR (1993) Perineuronal nets of extracellular matrix around parvalbumin-containing neurons of the hippocampus. Hippocampus 3:55–60.

Celio MR Blümcke I (1994) Perineuronal nets—a specialized form of extracellular matrix in the adult nervous system. Brain Res Rev 19:128–145.

Cepko C (1988) Retrovirus vectors and their applications in neurobiology. Neuron 1:435–453.

Cicero TJ Cowan WM Moore BW Suntzeff V (1970) The cellular localization of the two brain specific proteins S-100 and 14-3-2. Brain Res 18:25–34.

Dahl D Bignami A (1985) Intermediate filaments in nervous tissue, in Cell and Muscle Motility vol. 6 (Shay JW, ed.), Plenum, New York, pp. 75–96.

Dahl D Björklund H Bignami A (1986) Immunological markers in astrocytes, in Astrocytes (Fedoroff S Vernadakis A, eds.), Academic, New York, pp. 1–25.

Dahl D Chi NH Miles LE Nguyen BT Bignami A (1982) Glial fibrillary acidic (GFA) protein in Schwann cells: fact or artifact? J Histochem Cytochem 30:912–918.

DeAngelis PL Papaconstantinou J Weigel PH (1993) Molecular cloning, identification, and sequence of the hyaluronan synthase gene from group-A streptococcus-pyogenes. J Biol Chem 268:19181–19185.

Deeler P Nakanishi S (1983) Extracellular matrix distribution during neocortical wall ontogenesis in "normal" and "reeler" mice. J Hirnforsch 24: 209–224.

Edwards MA Yamamoto M Caviness VS Jr (1990) Organization of radial glia and related cells in the developing murine CNS. An analysis based upon a new monoclonal antibody marker. Neuroscience 36:121–144.

Fisher M (1984) Neuronal influence on glial enzyme expression: evidence from mutant mouse cerebella. Proc Natl Acad Sci USA 81:4414–4418.

Fisher M Gapp DA Kozak LP (1981) Immunohistochemical localization of sn-glycerol-3-phosphate dehydrogenase in Bergmann glia and oligodendroglia in the mouse cerebellum. Dev Brain Res 1:341–354.

Fryer HGL Kelly GM Molinaro L Hockfield S (1992) The high molecular weight Cat-301 chondroitin sulfate proteoglycan from brain is related to the large aggregating proteoglycan from cartilage aggrecan. J Biol Chem 267: 9874–9883.

Gadisseux J-F Evrard P (1985) Glial neuronal relationship in the developing central nervous system. A histochemical-electron microscope study of radial glial cell particulate glycogen in normal and reeler mice and the human fetus. Dev Neurosci 7:12–32.

Gard AL White FP Dutton GR (1985) Extra-neural glial fibrillary acidic protein (GFAP) immunoreactivity in perisinusoidal stellate cells of the rat liver. J Neuroimmunol 8:359–375.

Ghandour MS Labourdette G Vincendon G Gombos G (1981) A biochemical and immunohistological study of S100 protein in developing rat cerebellum. Dev Neurosci 4:98–109.

Giacobini E (1961) Localization of carbonic anhydrase in the nervous system. Science 134:1524–1525.

Girgrah N Letarte M Becker LE Cruz TF Theriaut E Moscarello MA (1991) Localization of the CD44 glycoprotein to fibrous astrocytes in normal white matter and to reactive astrocyte inactive lesions in multiple sclerosis. J Neuropath Exp Neurol 50:779–792.

Giulian D Baker TJ (1986) Characterization of ameboid microglia isolated from developing mammalian brain. J Neurosci 6:2163–2178.

Giulian D Chen J Ingeman JE George JK Noponen M (1989) The role of mononuclear phagocytes in wound healing after traumatic injury to adult mammalian brain. J Neurosci 9:4416–4429.

Golgi C (1903) Intorno alla struttura delle cellule nervose. Communicated to the Societá Medico-chirurgica of Pavia on April 19 1898, in Opera Omnia vol. 2 (Camillo Golgi, ed.), Hoepli, Milan, pp. 643–53.

Hardingham TE Fosang AJ (1992) Proteoglycans: many forms and many functions. Faseb J 6:861–870.

Hatfield JS Skoff RP Maisel H Eng LF (1984) Glial fibrillary acidic protein is localized in the lens epithelium. J Cell Biol 98:1895–1898.

Hatten ME (1990: Riding the glial monorail: a common mechanism for glial-guided neuronal migration in different regions of the developing mammalian brain. Trends Neurosci 13:179–183.

Held H (1902) Ueber den Bau der grauen und weissen. Substanz Arch Anat Psychol Abt V–VI:189–224.

His W (1889) Die Neuroblasten und deren Entstehung im embryonalen Mark. Arch Anat Physiol Anat Abt 251–300.

Hynes RO Patel R Miller RH (1986) Migration of neuroblasts along pre-existing axonal tracts during prenatal cerebellar development. J Neurosci 6:867–876.

Jessen KR Mirsky R (1980) Glial cells in the enteric nervous system contain glial fibrillary acidic protein. Nature 286:736–737.

Jessen KR Thorpe R Mirsky R (1984) Molecular identity distribution and heterogeneity of glial fibrillary acidic protein: an immunoblotting and immunohistochemical study of Schwann cells, satellite cells, enteric glia and astrocytes. J Neurocytol 13:187–200.

Juurlink BHJ Schousboe A Jørgensen OS Hertz L (1981) Induction by hydrocortisone of glutamine synthetase in mouse primary astrocyte cultures. J Neurochem 36:136–142.

Knudson CB (1993) Hyaluronan receptor-directed assembly of chondrocyte pericellular matrix. J Cell Biol 120:825–834.

Kumpulaien T Dahl D Korhonen KL Nyström SHM (1983) Immunolabeling of carbonic anhydrase isoenzyme C and glial fibrillary acidic protein in paraffin embedded tissue sections of human brain and retina. J Histochem Cytochem 31:879–886.

Lafarga M Berciano MT Blanco M (1984) The perineuronal net in the fastigial nucleus of the rat cerebellum. A Golgi and quantitative study. Anat Embryol 170:79–85.

Laurent TC Fraser JRE (1992) Hyaluronan. Faseb J 6:2397–2404.

Lendhahl U Zimmerman LB McKay RDG (1990) CNS stem cells express a new class of intermediate filament protein. Cell 60:585–595.

Levitt P Rakic P (1980) Immunoperoxidase localization of glial fibrillary acidic protein in radial glial cells and astrocytes of the developing rhesus monkey brain. J Comp Neurol 193(3)417–448.

Ling E-A Wong W-C (1993) The origin and nature of ramified and ameloid microglia: a historical review and current concepts. Glia 7:9–18.

Linnemann D Bock E (1989) Cell adhesion molecules in neural development. Dev Neurosci 11: 149–173.

Linser P Moscona AA (1979) Induction of glutamine synthetase in embryonic neural retina: localization in Müller fibers and dependence on cell interactions. Proc Natl Acad Sci USA 76:6476–6480.

Linser P Moscona AA (1983) Hormonal induction of glutamine synthetase in cultures of embryonic retina cells: requirement of neuron-glia contact interactions. Dev Biol 96:529–534.

Ludwin SK Kosek JC Eng LF (1976) The topographical distribution of S-100 and GFA proteins in the adult rat brain: an immunohistochemical study using horse-radish peroxidase-labeled antibodies. J Comp Neurol 165:197–208.

Luskin MB Pearlman AL Sanes JR (1988) Cell lineage in the cerebral cortex of the mouse studied in vivo and in vitro with a recombinant retrovirus. Neuron 1:635–647.

Lüth HJ Fischer J Celio MR (1992) Soybean lectin binding neurons in the visual cortex of the rat contain parvalbumin and are covered by glial nets. J Neurocytol 21:211–221.

Matrisian LM (1990) Metalloproteinases and their inhibitors in matrix remodeling. Trends Genet 6:121–125.

McCarthy KD de Vellis J (1980) Preparation of separate astroglial and oligoden-droglial cell cultures from rat cerebral tissue. J Cell Biol 85:890–902.

McGeer PL Kawamata T Walker DG Akiyama H Tooyama I McGeer EG (1993) Microglia in degenerative neurological disease. Glia 7:84–92.

McKay RDG Hockfield S (1982) Monoclonal antibodies distinguish antigeni-cally discrete neuronal types in the vertebrate central nervous system. Proc Natl Acad Sci USA 79:6747–6751.

Miller RH Fulton BP Raff MC (1989) A novel type of glial cell associated with nodes of Ranvier in rat optic nerve. Eur J Neurosci 1:172–180.

Mokuno K Kamholz J Behrman T Black C Sessa M Feinstein D Lee V Pleasure D (1989) Neuronal modulation of Schwann cell glial fibrillary acidic pro-tein (GFAP). J Neurosci Res 23:396–405.

Moore BW (1969) Acidic proteins, in Handbook of Neurochemistry vol. I Chemical architecture of the nervous system (Lajtha A, ed.), Plenum, New York, pp. 93–99.

Nicholson C Rice ME (1986) The migration of substances in the neuronal microenvironment, in The Neuronal Microenvironment (Cserr HF ed.), Ann NY Acad Sci 481:55–68.

Noble M Murray K Stroobant P Waterfield MD Ruddle P (1988) Platelet-derived growth factor promotes division and motility and inhibits premature differentiation of the oligodendrocyte/type-2 astrocyte progenitor cell. Nature 280:688–690.

Norenberg MD (1979) The distribution of glutamine synthetase in the rat cen-tral nervous system. J Histochem Cytochem 27:756–762.

Norenberg MD (1983) Immunohistochemistry of glutamine synthetase, in Gluta-mine Glutamate and GABA in the Central Nervous System (Hertz KL Kvamme E McGeer EG Schousboe A, eds.), Liss, New York, pp. 95–111.

Norenberg MD Dutt K Reif-Lehrer L (1980) Glutamine synthetase localization in cortisol-induced chick embryo retinas. J Cell Biol 84:803.

Palladini G Alfei L (1968) Histochimie du neuropile du poulet pendant l'ontogénèse—hyperstrié vrai. Histochemie 14:314–323.

Penfield W (1932) Cytology and cellular pathology of the nervous system (Penfield W, ed.), Hafrer, New York, reprinted in 1965.

Perides G Lane WS Andrews D Dahl D Bignami A (1989) Isolation and partial characterization of a glial hyaluronate-binding protein. J Biol Chem 264:5981–5987.

Perides G Rahemtulla F Lane WS Asher RA Bignami A (1992) Isolation of a large aggregating proteoglycan from human brain. J Biol Chem 267:23883–23887.

Perides G Erickson HP Rahemtulla F Bignami A (1993) Colocalization of tenascin, with versican, a hyaluronate-binding chondroitin sulfate proteoglycan. Anat Embryol 188:467–479.

Perry VH Matyszak MK Fearn S (1993) Altered antigen expression of microglia in the aged rodent CNS. Glia 7:60–67.

Prehm P (1986) Mechanism localization and inhibition of hyaluronate synthesis, in Articular Cartilage Biochemistry (Kuettner KE Schleyerbach R Hascall VC, eds.), Raven, New York, pp. 81–89.

Price J Thurlow L (1988) Cell lineage in the rat cerebral cortex: a study using retroviral-mediated gene transfer. Development 104:473–482.

Raff MC Miller RH Noble M (1983) A glial progenitor cell that develops *in vitro* into an astrocyte or an oligodendrocyte depending on culture medium. Nature 303:390–396.

Raff MC Mirsky R Fields KL Lisak RP Dorfman SH Silverberg DH Gregson NA Leibowitz S Kennedy MC (1978) Galactocerebroside is a specific cell-surface antigenic marker for oligodendrocytes in culture. Nature 274:813–816.

Raff MC Lillien LE Richardson WD Burne JF Noble MD (1988) Platelet-derived growth factor from astrocytes drives the clock that times oligodendrocyte development in culture. Nature 333:562–564.

Richardson WD Pringle N Mosley MJ Westermark B Dubdis-Dalcq M (1988) A role for platelet-derived growth factor in normal gliogenesis in the central nervous system. Cell 53:309–319.

Ripellino JA Klinger MM Margolis RU Margolis RK (1985) The hyaluronic acid binding region as a specific probe for the localization of hyaluronic acid in tissue sections. Application to chick embryo and rat brain. J Histochem Cytochem 33:1060–1066.

Roback HN Scherer HJ (1935) Über die feinere Morphologie des frühkindlichen Hirnes unter besonderer Berücksichtigung der Gliaentwicklung Virchows. Arch Pathol Anat 294:339–354.

Sanes JR Rubenstein JLR Nicolas J-F (1986) Use of a recombinant retrovirus to study postimplantation cell lineage in the mouse. EMBO J 5: 3133–3142.

Schaper A (1897) Die frühesten Differenzierungsvorgänge in Central nerven system. Arch Entwickl-Mech Org 5:81–132.

Schmechel DE Rakic P (1979) A Golgi study of radial glial cells in developing monkey telencephalon: morphogenesis and transformation into astrocytes. Anat Embryol 156:115–152.

Schnitzer J (1988) Astrocytes in the guinea pig, horse, and monkey retina: their occurrence coincides with the presence of blood vessels. Glia 1:74–89.

Scott JE (1985) Proteoglycan histochemistry. A valuable tool for connective tissue biochemists. Collagen Rel Res 5:541–575.

Scott JE Heatley F Hull WE (1984) Secondary structure of hyaluronate in solution. A ^{1}H-n.m.r investigation at 300 and 500 MH$_2$ in [^{2}H$_6$] dimethyl sulphoxide solution. Biochem J 220:197–205.

Sedgwick JD Schwender S Imrich H Dorries R Butcher GW (1991) Isolation and direct characterization of resident microglial cells from the normal and inflamed central nervous system. Proc Natl Acad Sci USA 88:7438–7442.

Shinomura T Nishida Y Kimata K (1992) A large chondroitin sulfate proteoglycan (PG-M) and cartilage differentiation, in Articular Cartilage and Osteoarthritis (Kuettner K et al. eds.), Raven, New York, pp. 35–44.

Sidman RL Rakic P (1973) Neuronal migration with special reference to developing human brain: a review. Brain Res 62:1–35.

Skoff RP (1990) Gliogenesis in rat optic nerve: Astrocytes are generated in a single wave before oligodendrocytes. Develop Biol 139:149–168.

Stensaas LJ Brugess PR Horch KW (1979) Regenerating dorsal root axons are blocked by spinal cord astrocytes. Soc Neurosci Abstr 5:684.

Stensaas LJ Partlow LM Burgess PR Horch KW (1987) Inhibition of regeneration: the ultrastructure of reactive astrocytes and abortive axon terminals in the transitional zone of the dorsal root. Prog Brain Res 71:457–468.

Takahashi T Misson J-P Caviness VS (1990) Glial process elongation and branching in the developing murine neocortex: a qualitative and quantitative immunohistochemical analysis. J Comp Neurol 302:15–28.

Toole BP (1991) Proteoglycans and hyaluronan in morphogenesis and differentiation, in Cell Biology of Extracellular Matrix (Hay ED, ed.), Plenum, New York, pp. 305–341.

Trotter J Schachner M (1989) Cells positive for the O4 surface antigen isolated by cell sorting are able to differentiate into astrocytes or oligodendrocytes. Dev Brain Res 46:115–122.

Trowbridge IS (1991) A prototype for transmembrane protein tyrosine phosphatases. J Biol Chem 266:23517–23520.

Turner DL Cepko CL (1987) A common progenitor for neurons and glia persists in rat retina late in development. Nature 328:131–136.

Underhill C (1992) CD44-The hyaluronan receptor. J Cell Sci 103:293–98.

Underhill CB Toole BP (1979) Binding of hyaluronate to the surface of cultured cells. J Cell Biol 82:475–484.

Underhill CB Thurn AL Lacy BE (1985) Characterization and identification of the hyaluronate-binding site from membranes of SV-ST3 cells. J Biol Chem 260:8128–8133.

Ushkaryov YA Petreuko AG Geppert M Sudhof TC (1992) Neuronexins: synaptic cell surface proteins related to the alpha-latrotoxin receptor and laminin. Science 257:50–56.

Van Harreveld A Crowell J Malhotra SK (1965) A study of extracellular space in central nervous tissue by freeze-substitution. J Cell Biol 25:117–137.

Vaysse PJ-J Goldman JE (1990) A clonal analysis of glial lineages in neonatal forebrain development *in vitro*. Neuron 5(3)227–235.

Vernadakis A Woodbury DM (1965) Cellular and extracellular spaces in developing brain. Arch Neurol 12:284–293.

Vogel H Butcher EC Picker LJ (1992) H-CAM expression in the human nervous system: evidence for a role in diverse glial interactions. J Neurocytol 21:363–373.

Walsh C Cepko CL (1992) Widespread dispersion of neuronal clones across functional regions of the cerebral cortex. Science 255:434–440.

Watanabe T Raff MC (1988) Retinal astrocytes are immigrants from the optic nerve. Nature 332:834–837.

Weigert C (1895) Beiträge zur Kenntnis der normalen menschlichen Neuroglia. Moritz Diesterweg, Frankfurt A.M.

Yang H-Y Lieska N Shao D Kriho V Pappas GD (1993) Immunotyping of radial glia and their glial derivatives during development of the rat spinal cord. J Neurocytol 22:558–571.

Yeh H-J Silos-Santiago I Wang YX George RJ Snider WD Deuel TF (1993) Developmental expression of the platelet derived growth factor α-receptor gene in mammalian central nervous system. Proc Natl Acad Sci 90:1952–1956.

Yen S-H Fields KL (1981) Antibodies to neurofilament glial filament and fibroblast intermediate filament proteins bind to different cell types of the nervous system. J Cell Biol 88:115–126.

Zimmermann DR Ruoslahti E (1989) Multiple domains of the large fibroblast proteoglycan, versican. EMBO J 8:2975–2981.

Evolution of Astrocytes in the Vertebrate CNS

Isabel Suárez, Guillermo Bodega,
Miguel Rubio, and Benjamin Fernández

1. Introduction

It has become more and more clear that astrocytes can no longer be considered as a homogeneous cell population. Astrocytes are heterogeneous in different brain regions and at different developmental stages, as well as in different vertebrate groups.

Astrocyte morphological differentiation and cell process outgrowth depend on microtubules and on gliofilament assembly. Vimentin and glial fibrillary acidic protein (GFAP) represent the principal constituents of the intermediate filaments found in astrocytes (Dahl, 1981). In the adult brain, the gliofilaments are mainly composed of GFAP (Roots, 1982; Eng, 1985). Thus, GFAP immunostaining is the most reliable way of identifying astrocytes. The demonstration that the organization and accumulation of gliofilaments increase in parallel to the concentration of GFAP during in vitro astrocytes differentiation (Sensenbrenner et al., 1980) supports this application.

The distribution of GFAP has been studied in the developing and the adult mammalian brain (Dupouey et al., 1985; Suárez et al., 1987; Hirano and Goldman, 1988; McDermott and Lantos, 1989), as well as in other vertebrates (Onteniente et al., 1983; Dahl et al., 1985; Roots, 1986; Nona et al., 1989; Cardone and Roots, 1990; Monzón-Mayor et al., 1990; Cameron-Curry et al., 1991). It has been postulated that the presence of GFAP in submammals is extremely constant in so far as its biochemical properties (Dahl

From: *Neuron-Glia Interrelations During Phylogeny: I. Phylogeny and Ontogeny of Glial Cells*
A. Vernadakis and B. Roots, Eds. Humana Press Inc., Totowa, NJ

41

and Bignami, 1973), although morphological and topographical differences have been observed between the species. Therefore, because of such a high degree of stability during evolution, GFAP might be a valuable tool in phylogenetical studies on astroglial evolution.

With the use of classic metallic impregnations (Cajal's gold sublimate and Golgi-Rio Hortega methods) and immunohistochemical (anti-GFAP and antivimentin) methods, we have identified different astrocyte types in a variety of vertebrate models on the basis of their morphology and distribution. This chapter is based on published and unpublished studies of in vivo astrocytes types to determine the possible phylogenetic evolution of astrocytes in the vertebrate scale. Here, the term "astrocyte" refers to cells that stained with classic techniques and also were immunoreactive for the GFAP antibody. Their morphologies varied with the technique used, the type of astrocytes, the cell location, and the vertebrate group.

2. Ependymal Astrocytes

The ependyma is composed of different types of ependymal cells, and its composition varies not only in different CNS regions (Bruni et al., 1985), but also in the same region of the CNS when different vertebrate groups are compared (Ramón y Cajal, 1919; Horstmann, 1954). The term "ependymal astrocytes" was introduced by Hortsmann (1959).

Ependymocytes that express GFAP can be considered as astrocytes. The number of ependymal cells that maintain GFAP expression in adult animals and their astroglial character are very different in the vertebrate scale. Therefore, the astroglial nature of ependymal cells is, in many cases, a transient characteristic.

Glial fibrillary acidic protein-positive ependymocytes have been observed in the fish CNS (Anderson et al., 1984; Nona et al., 1989; Cardone and Roots, 1990; Rubio et al., 1992; Bodega et al., 1992, 1993); they were abundant in the III ventricle (Fig. 1A) and the strongest GFAP immunoreactivity was located in the somata and in the proximal region of the radial process (Fig. 1B). Most of the GFAP-positive ependymal astrocytes possessed a tanycytic character demonstrated by studying serial sections; the presence of tanycytes was also demonstrated using metallic impregnations (Fig. 1C).

The results obtained in the literature on GFAP expression in the amphibian CNS are inconsistent. GFAP-negative ependymocytes (Onteniente et al., 1983, Zamora and Mutin, 1988) or with low GFAP expression (Miller and Liuzzi, 1986; Bodega et al., 1990a) have been mentioned. The poor GFAP expression in this vertebrate group has also been demonstrated by means of immunoblotting techniques (Dahl and Bignami, 1973; Dahl et al., 1985).

The distribution of GFAP-positive ependymocytes in the reptilian CNS is very heterogeneous (Bodega et al., 1990b; Monzón-Mayor et al., 1990; Yanes et al., 1990). A large number of GFAP-positive ependymal astrocytes were observed in the lizard diencephalon (Fig. 1D) and they showed a very strong immunoreaction in their cell body and in the long and straight radial process (Fig. 1E,F), which frequently reached blood vessels (Fig. 1F).

No GFAP expression has been demonstrated in ependymocytes in adult birds (Onteniente et al., 1983, Alvarez-Buylla et al., 1987, Cameron-Curry et al., 1991, Kálmán et al., 1993, Bodega et al., 1994).

The number of GFAP-positive ependymocytes varies in the ontogenetic development of mammals (Roessmann et al., 1980; Bruni et al., 1985; Suárez et al., 1987; Sarnat 1992) and a lack of GFAP expression in ependymocytes from different adult CNS regions has also been described (Onteniente et al., 1983; Bullón et al., 1984; Didier et al., 1986; Hajós and Kálmán 1989). However, GFAP-positive ependymal astrocytes were seen in the III ventricle (Fig. 1G,H) as has already been described (Bascó et al., 1981, Suárez et al., 1987), and they participate in the formation of perivascular sheaths (Fig. 1H). The few GFAP-positive ependymocytes described show a tanycytic character (Levitt and Rakic, 1980; Roessmann et al., 1980; Bascó et al., 1981; Didier et al., 1986; Leonhardt et al., 1987).

3. Radial Glia

Radial glia are mainly located near the ventricular layer and they have long radial processes, that come into contact with the external limiting membrane. The term "radial glia" was reintroduced by Rakic (1981) and in the literature frequently comprises not only radial glia but also ependymal astrocytes and isolated radial processes.

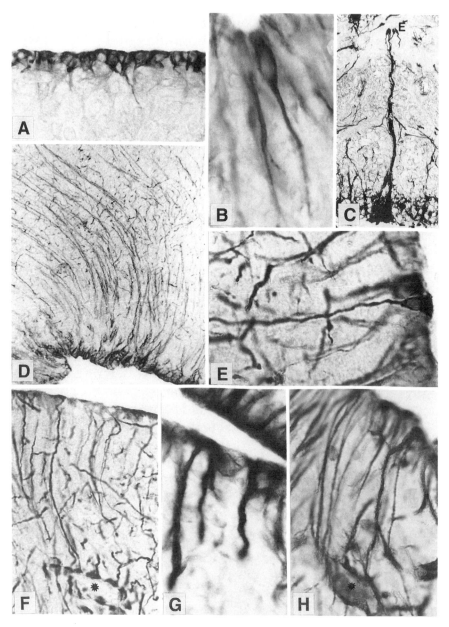

Fig. 1. **(A)** GFAP-immunoreactive ependymocytes. Fish hypothalamus. **(B)** GFAP-positive ependymocyte. Fish. **(C)** Ependymocytes (E) crossing the spinal cord. Fish. Golgi-Río Hortega method. **(D)** GFAP-positive ependymocytes in the lizard cerebral cortex. **(E)** GFAP-immunoreactive ependymocytes of lizard hypothalamus. **(F)** GFAP-positive tanycyte

The existence of GFAP-positive radial glia in the fish CNS has been described by numerous authors (Anderson et al., 1984; Nona et al., 1989; Cardone and Roots, 1990; Bodega et al., 1993). Although the somata of the radial glia usually are not immunohistochemically recognizable, they have recently been reported in the rombencephallon (Rubio et al., 1992), as well as in proximity to the ventricles (Fig. 2A). Using metallic impregnations, some radial glia are easily recognizable even when they were far from the ependyma; these had a pear-shaped somata with numerous short irregular processes radiating from the cell bodies, and a long radial process that reached the subpial glia limitans (Fig. 2B).

The glioarchitecture of amphibians is similar to that in fish. The radial glial cell is the only nonependymal astrocyte in the amphibian CNS (Ramón y Cajal, 1909–1911; Miller and Liuzzi, 1986; Zamora and Mutin, 1988; Bodega et al., 1990a).

Radial glia could be demonstrated in the CNS of adult reptiles by means of immunohistochemical techniques (Fig. 2C–E) and metallic impregnations (Fig. 2F), and their morphological characteristics were similar to those of fish and amphibians. These GFAP-positive cells were scarce and the amount varied in the different regions of the CNS; for example, they were more abundant in the hypothalamus than in the spinal cord.

Although Alvarez-Buylla et al. (1987) did not find GFAP-positive radial cells, Kálmán et al. (1993) have recently demonstrated the existence of a well-developed GFAP-positive radial network in the adult bird brain.

Radial glia were also observed in adult mammals, mainly located in the diencephalon (Fig. 2G). They have also been described in the spinal cord (Levitt and Rakic, 1980; Bodega et al., 1994). At present, it is not possible to know whether the elements called radial glia-like cells (Bonfanti et al., 1993), radially oriented

---

related to blood vessel (*). Lizard. **(G)** GFAP immunoreactive ependymocytes in the III ventricle of the mammalian hypothalamus. **(H)** GFAP-positive tanycytes in the arcuate nucleus related to blood vessel (*). Hamster hypothalamus. Compare the different immunoreactivity between reptilian (in F) and mammalian (in H) tanycytes. Magnification rates: (A) ×375; (B) ×1.156; (C) ×13.5; (D) ×188; (E) ×804; (F) ×385; (G) ×1005; (H) ×686.

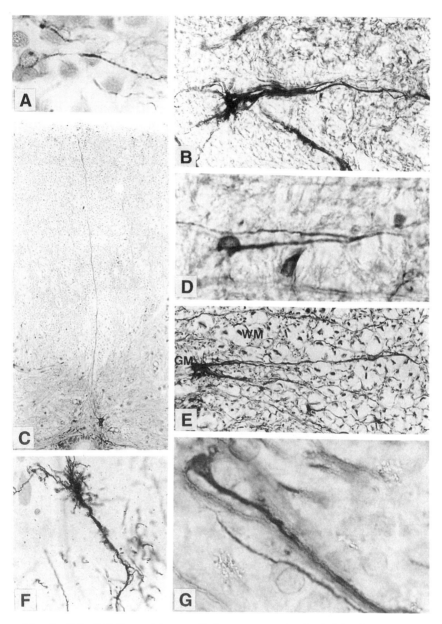

Fig. 2. **(A)** GFAP-positive radial astrocyte. Fish. **(B)** Fish radial glia. Golgi-Río Hortega method. **(C)** Reptilian GFAP-positive radial glia crossing the spinal cord. **(D)** GFAP-immunoreactive radial glia in the lizard cerebral cortex. **(E)** Radial glia in the lizard spinal cord, whose somata is located in the limit between grey (GM) and white matter (WM).

astrocytes (Liuzzi and Miller 1987), and cells with radial processes (Reichenbach 1990), which have been described in the adult mammal CNS, are actually radial glia.

In addition to the preceding cells, the adult CNS also contains some astrocytes that maintain the radial arrangement: Bergmann glial cells in the cerebellum (Bignami and Dahl, 1974; Bovolenta et al., 1984; Suárez et al., 1992) and Müller cells in the retina (Bignami and Dahl, 1979; Björklund et al., 1985).

4. Astrocytes

Astrocytes were originally named for the star shape created by their radiating processes (Ramón y Cajal, 1909–1911). Studies with classic metallic methods demonstrated two morphologically distinct types of astrocytes in the CNS of mammals: protoplasmic astrocytes in the grey matter and fibrous astrocytes in the white matter. This early classification was confirmed by electron microscopy and astrocytes are characterized by a large number of intermediate filaments, which are even more numerous in fibrous than in protoplasmic astrocytes (Peters et al., 1976). Since GFAP is the main protein in intermediate filaments of differentiated astrocytes (Eng, 1985), "fibrous" and "protoplasmic" astrocytes present different amounts of GFAP.

The most primitive vertebrate group with astrocytes are reptiles (Ramón y Cajal, 1909–1911; King, 1966; Stensaas and Stensaas, 1968; Batista et al., 1981). Although these cells are GFAP-positive (Onteniente et al., 1983; Bodega et al., 1990b), the number is quite reduced when compared to higher vertebrates.

Astrocyte shape and number of astrocytic processes vary characteristically with the technique, the different brain area, the age, and the vertebrate group (Fig. 3A–E). The distribution of astrocytes is not random, it is determined by the types of neurons present and also by the association with synapses (Suárez et al., 1980; Fernández et al., 1984; García-Segura et al., 1986; Vernadakis,

---

(F) Radial glia in the lizard cerebral cortex. Golgi-Río Hortega method. (G) GFAP-positive radial astrocyte in the mammalian hypothalamus. Magnification rates: (A) ×553; (B) ×430; (C) ×160; (D) ×1005; (E) ×368; (F) ×368; (G) ×1005.

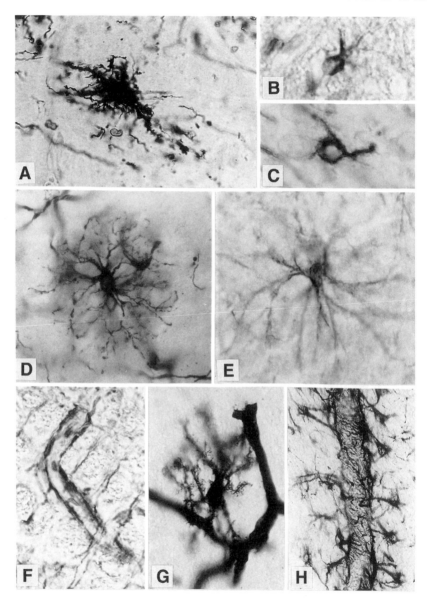

Fig. 3. **(A)** Reptilian astrocyte. Golgi-Río Hortega method. **(B)** GFAP-positive astrocyte in the adult lizard cerebral cortex. **(C)** GFAP-positive astrocyte in the developing mammalian hypothalamus, the immunoreactive processes of the immature astrocyte can be compared to the adult reptilian astrocytic processes (in B). **(D)** Diencephalic mammalian astrocyte. Golgi-Río Hortega method. **(E)** GFAP-

1986; Meshul et al., 1987), Ranvier nodes (Raine, 1984; Black and Waxman, 1988; Suárez and Raff, 1989), and by the presence of blood vessels, ventricles, and external brain surfaces (Suárez and Fernández, 1983; Fernández et al., 1984; Suárez et al., 1994). Astrocytes and/or their processes form the perivascular (Fig. 3F–H), periependymal (Fig. 4A,B), and subpial (Fig. 4C–E) limiting membranes. Although the mammalian blood vessels were surrounded by a complete perivascular astroglial sheath (Figs. 3H and 4G), not all vessels are completely covered by astroglial processes in submammalian vertebrates. In birds, only a few blood vessels are surrounded by perivascular GFAP-positive astrocytes (Alvarez-Buylla et al., 1987), whereas the few GFAP-positive astrocytes in reptiles are mostly related to blood vessels (Fig. 4C,F); reptilian astrocytes may send long processes that, joined to the radial glial processes, form the subpial glia limitans (Fig. 4C).

Ramón y Cajal (1909–1911) suggested that radial glial cells were precursors of astrocytes, as was demonstrated almost a century later (Levitt and Rakic 1980, Benjelloun-Touimi et al., 1985; Voight, 1989). A series of transitional forms, from the bipolar radial glial cell to the multipolar astrocyte, can be identified with a monoclonal antibody that recognizes radial glial cells (Misson et al., 1988). In the adult brain of reptiles, birds, and mammals, it is possible to observe intermediate forms between astrocytes and radial glial cells (Gianonatti et al., 1984; Suárez et al., 1987; Bodega et al., 1990b); some of these intermediate forms in the mammalian brain could be considered remnants of primitive radial glia (Rakic, 1981). It is important to remember, however, that not all the radial glia is replaced in the adult higher vertebrates by stellate astrocytes, as seems to have been the case for tanycytes and Bergmann glial cells.

---

immunopositive mammalian astrocyte. **(F)** Reptilian blood vessel surrounded by GFAP-positive radial astrocytic processes. **(G)** Mammalian astrocyte, whose processes are related to two blood vessels. Golgi-Río Hortega method. **(H)** Mammalian blood vessel completely surrounded by GFAP-positive astrocytes. Rat cerebellum. Magnification rates: (A) ×335; (B) ×804; (C) ×804; (D) ×395; (E) ×804; (F) ×804; (G) ×335; (H) ×245.

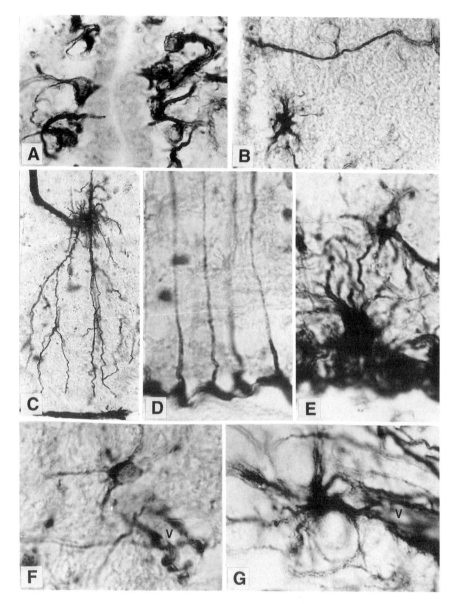

Fig. 4. (A) GFAP immunoreactive periependymal astrocytes in the adult mammalian hypothalamus. (B) Periependymal astrocyte and ependymocyte can coexist in the adult mammalian hypothalamus. (C) Perivascular astrocyte sending long processes to form the sub-pial limitans in the lizard spinal cord. Golgi-Río Hortega. (D) The subpial limitans in the lizard cerebral cortex consists of GFAP-positive radial astroglial processes. (E) The subpial glial limitans in the mam-

5. Conclusions

Studies on the organization of astrocytes provide evidence of heterogeneity within the CNS that may correlate with different functions. The astroglial character of ependymal astrocytes, radial glia, and astrocytes has been confirmed by their GFAP immunoreactivity.

It has been reported that the predominant type of astrocytes in fish, amphibians, and reptiles is the ependymal astrocyte (King, 1966; Roots, 1986). However, our own observations demonstrate that the ependymal astrocyte is the predominant cell type in the entire fish CNS and in many parts of the amphibian CNS, but its distribution is restricted to certain brain areas in reptiles. Glial fibrillary acidic protein expression in ependymal cells is inversely related with phylogenetical evolution; the number of GFAP-positive ependymocytes is higher in lower vertebrates than in higher vertebrates, although all the ependymal cells in higher vertebrates are GFAP-positive during the first phases of development (Levitt and Rakic, 1980; Roessmann et al., 1980; Suárez et al., 1987; Gould et al., 1990; Sarnat, 1992). Considering that most GFAP-positive ependymocytes in the adult vertebrate CNS have a tanycytic character, it would be interesting to know the precise relationships between tanycytes and common ependymal cells. Gould et al., (1990) have suggested that mammal ependymocytes could have originated from tanycytes and a loss of GFAP would accompany this maturation process.

The astrocyte character of radial glia has also been immuno-histochemically demonstrated. The radial glia is the astrocyte that is present in all vertebrate groups; radial glia are abundant in fish, predominant in amphibians and very scarce in reptiles, birds, and mammals. The similarities between radial glia in lower vertebrates and the radial glia observed in the immature brain of higher vertebrates suggests that the radial orientation is

---

malian hypothalamus consists of astrocytes and astrocytic processes. GFAP immunoreactivity. **(F)** Reptilian astrocyte closes to cerebral blood vessel (V). GFAP immunoreactivity. **(G)** Note the higher GFAP immunoreactivity in the mammalian perivascular astrocyte when compared to reptilian astrocyte in F. Blood vessel (V). Magnification rates: (A) ×703; (B) ×650; (C) ×240; (D) ×1070; (E) ×1140; (F) ×1206; (G) ×1070.

phylogenetically retained during evolution. During phylo-
genesis, the transformation of these GFAP-positive radial glia
into astrocytes has been demonstrated in mammals (Levitt and
Rakic, 1980; Suárez et al., 1987; Choi, 1988), with concomitant
changes in structure.

The appearance of astrocytes in reptiles determines the
decrease of radial glia as well as ependymal astrocytes, suggesting a cellular substitution. Therefore, reptiles are considered the
key group in the phylogenetical evolution of astrocytes (Bodega
et al., 1990b), because it is the first vertebrate group in which
astrocytes appear. The three astrocyte types, ependymal astrocytes,
radial glia, and astrocytes, are simultaneously present from reptiles on.

In this phylogenetic evolution, the astrocytes progressively
separate from the ependymal layer, shorten their radial processes,
and increase the number of star processes. Therefore, the radially
organized structures disappear in phylogeny concomitant with
astroglial differentiation. With the increase in the number and differentiation of astrocytes, specialization also increases and, thus,
potential is lost. This apparent loss of potential is recovered after
injury and/or in pathological processes.

The astrocytes in lower vertebrates have to cover all the functions ascribed to astrocytes in higher vertebrates. In lower vertebrates, the radial processes are regionally specialized (Miller and
Liuzzi, 1986), whereas in higher vertebrates there is an astrocyte
specialization.

An increase in the relationships of astrocytes to blood vessels and pial zones has been observed in the phylogenetic scale
(Achúcarro, 1913; King, 1966; Bodega et al., 1990b), as well as
during the ontogenetic development in mammals (Suárez et al.,
1987). Both the glial perivascular and subpial limitans consist of
ependymal and radial glia in lower vertebrates, astrocytes are
incorporated in reptiles, and in adult higher vertebrates both
limitans are formed by astrocytes. The fine cover of both the
perivascular and the subpial limitans and the presence of radial
glia during brain maturation in mammals correspond to characteristics found in the adult brain of lower vertebrates.

Phylogenetic variations may also exist in the membrane of
astrocytes. Orthogonal arrays are characteristic elements of mammalian astrocytes, whereas they are virtually absent of astrocytes

in lower vertebrates (Korte and Rosenbluth, 1981; Wolburg et al., 1983; Wujek and Reier, 1984), as well as in the plasma membranes of immature astrocytes in the brain of pre- and neonatal rats (Anders and Brightman, 1979).

It is a general rule that parts of the nervous system that may have appeared first in phylogeny have a tendency to appear early in ontogeny, and structures that are thought to have arisen later in evolution also often arise late in ontogeny. This rule was applied to neurons or brain regions (Angevine, 1970), but it has not been applied to astrocytes. The data shown in this chapter might suggest that not only the phylogenetically older neurons develop first ontogenetically, but also that the astrocytes do.

In summary, the phylogenetic development of astrocytes in the vertebrate brain resembles the ontogenetic development of the higher vertebrate CNS: Ependymocytes are the first cells with an astrocyte character, the next to appear are the radial glia, which are the precursors of astrocytes. However, in their evolution not all the GFAP-positive radial glia change the morphology from radial glial cells to star- shaped astrocytes; some ependymal astrocytes and radial glia restricted to certain brain areas remain in higher vertebrates, maintaining the morphologies present in lower vertebrates, and, probably, their functions.

Acknowledgments

This chapter was partially supported by grants from the DGICYT PB91-0167-C02-01 and from the UAH 93/5. The authors thank CF Warren (ICE, UAH) for her linguistic assistance.

References

Achúcarro N (1913) De l'evolution de la néuroglie, et spécialement de ses relations avec l'appareil vasculaire. Trab Lab Inv Biol 11:169–213.
Alvarez-Buylla A Burskirk DR Nottebohm F (1987) Monoclonal antibody reveals radial glia in adult avian brain. J Comp Neurol 264:159–170.
Anders JJ Brightman MW (1979) Assemblies of particles in the cell membranes of developing, mature and reactive astrocytes. J Neurocytol 8:777–795.
Anderson MJ Swanson KA Waxman SG Eng LF (1984) Glial fibrillary acidic protein in regenerating teleost spinal cord. J Histochem Cytochem 31:1099–1106.
Angevine JB (1970) Time of neuron origin in the diencephalon of the mouse. J Comp Neurol 139:129–187.

Bascó E Woodhams PL Hajós F Balázs R (1981) Immunocytochemical demonstration of glial fibrillary acidic protein mouse tanycytes. Anat Embryol 162:217–222.

Batista MAP Fernández B Suárez I (1981) Estudio de los componentes astrocitarios del hipotálamo del Chalcides viridanus (reptil, Scincidae). Morp Norm Patol 5:47–53.

Benjelloun-Touimi S Jacque CM Derer P De Vitry F Maunory R Dupouey P (1985) Evidence that mouse astrocytes may be derived from the radial glia. An immunohistochemical study of the cerebellum in the normal and reeler mouse. J Neuroimmunol 9:87–97.

Bignami A Dahl D (1974) Astrocyte-specific protein and neuroglial differentiation. An immunofluorescence study with antibodies to the GFAP. J Comp Neurol 153:27–38.

Bignami A Dahl D (1979) The radial glia of Müller in the rat retina and their response to injury. An immunofluorescence study with antibodies to the glial fibrillary acidic protein. Exp Eye Res 28:63–69.

Björklund H Eriksdotter-Nilsson M Dahl D Hoffer B Olson L (1985) Image analysis of GFAP-positive astrocytes from adolescence to senescence. Exp Brain Res 58:163–170.

Black JA Waxman SG (1988) The perinodal astrocyte. Glia 1:169–183.

Bodega G Suárez I Fernández B (1990a) Radial astrocytes and ependymocytes in the spinal cord of the adult toad (Bufo bufo L.). An immunohistochemical and ultrastructural study. Cell Tissue Res 260:307–314.

Bodega G Suárez I Rubio M Fernández B (1990b) Distribution and characteristics of the different astroglial cell types in the adult lizard (Lacerta lepida) spinal cord. Anat Embryol 181:567–575.

Bodega G Suárez I Rubio M Villalba RM Fernández B (1992) Hyperammonemia induces transient GFAP immunoreactivity changes in goldfish spinal cord (Carassius auratus L.). Neurosci Res 13:217–225 .

Bodega G Suárez I Rubio M Villaba RM Fernández B (1993) Astroglial pattern in the spinal cord of the adult barbel (Barbus comiza). Anat Embryol 187:385–395.

Bodega G Suárez I Rubio M Fernández B (1994) Ependyma: phylogenetical evolution of GFAP and vimentin expression in vertebrate spinal cord. Histochemistry 102:113–122.

Bonfanti L Poulain DA Theodosis DT (1993) Radial glia-like cells in the supraoptic nucleus of the adult rat. J Neuroendoc 5:1–5.

Bovolenta P Liem RKH Mason CA (1984) Development of cerebellar astroglia: transitions in form and cytoskeletal content. Dev Biol 102:248–259.

Bruni JE Bigio MR Clattenburg RE (1985) Ependyma: normal and pathological. A review of the literature. Brain Res Rev 9:1–19.

Bullón MM Alvarez-Gago T Fernández B Aguirre C (1984) Glial fibrillary acidic (GFAP) protein in rat spinal cord. An immunoperoxidase study in semithin sections. Brain Res 309:79–83.

Cameron-Curry P Aste N Viglietti-Panzica C Panzica GC (1991) Immunocytochemical distribution of glial fibrillary acidic protein in the central nervous system of the japanese quail (Coturnix coturnix japonica). Anat Embryol 184:571–581.

Cardone B Roots BI (1990) Comparative immunohistochemical study of glial filament proteins (glial fibrillary acidic protein and vimentin) in goldfish, octopus and snail. Glia 3:180–192.

Choi B (1988) Prenatal gliogenesis in the developing cerebrum of the mouse. Glia 1:308–316.

Dahl D (1981) The vimentin-GFA protein transition in rat neuroglia cytoskeleton occurs at the time of myelination. J Neurosci Res 6: 741–748.

Dahl D Bignami A (1973) Immunochemical and immuno-fluorescence studies of the GFAP in vertebrates. Brain Res 61:279–293.

Dahl D Crosby CJ Sethi JS Bignami A (1985) Glial fibrillary acidic (GFA) protein in vertebrates: immunofluorescence and immunoblotting study with monoclonal and polyclonal antibodies. J Comp Neurol 239:75–88.

Didier M Harandi M Aguera M Bancel B Tardy M Fages C Calas A Stagaard M Mollgard K Belin MF (1986) Differential immunocytochemical staining for GFA protein, S-100 protein and glutamine synthetase in the rat subcommissural organ, nonspecialized ventricular ependyma and adjacent neuropil. Cell Tissue Res 245:343–351.

Dupouey P Benjelloun S Gomes D (1985) Immunohistochemical demonstration of an organized cytoarchitecture of the radial glia in the CNS of the embryonic mouse. Dev Neurosci 7:81–93.

Eng LF (1985) Glial fibrillary acidic protein (GFAP) the major protein of glial intermediate filaments in differentiated astrocytes. J Neuroimmunol 8:203–214.

Fernández B Suárez I González G (1984) Topographical distribution of the astrocytic lamellae in the hypothalamus. Anat Anz 156:31–37.

García-Segura LM Baetens D Naftolin F (1986) Synaptic remodelling in arcuate nucleus after injection of estradiol valerate in adult female rats. Brain Res 366:131–136.

Gianonatti C Fernández B Suárez I Bodega G (1984) Glioarchitecture du cervelet de poulet. Arch Biol 95:71–82.

Gould SJ Howard S Papadaki L (1990) The development of ependyma in the human fetal brain: an immunohistological and electron microscopic study. Dev Brain Res 55:255–267.

Hajós F Kálmán M (1989) Distribution of glial fibrillary acidic protein (GFAP)-immunoreactive astrocytes in the rat brain. Exp Brain Res 78:164–173.

Hirano M Goldman JE (1988) Gliogenesis in rat spinal cord: evidence for origin of astrocytes and oligodendrocytes from radial precursors. J Neurosci Res 21:155–167.

Horstmann E (1954) Die faserglia des selachiergehirns. Z Zellforsch 39: 588–617.

Horstmann E (1959) Zur frage des extracellulären raumes in zentralnervensystem. Anat Anz 105:100–106.

Kálmán M Székely AD Csillag A (1993) Distribution of glial fibrillary acidic protein-immunopositive structures in the brain of the domestic chicken (*Gallus domesticus*). J Comp Neurol 330:221–237.

King JS (1966) A comparative investigation of neuroglia in representative vertebrates: a silver carbonate study. J Morphol 119:435–466.

Korte GE Rosenbluth J (1981) Ependymal astrocytes in the frog cerebellum. Anat Rec 199:267–279.

Leonhardt H Krisch B Erhardt H (1987) Organization of the neuroglia in the midsagittal plane of the central nervous system: a speculative report, in Functional morphology of Neuroendocrine Systems (Scharrer B Korf HW Hartwig HG, eds.), Springer-Verlag, Berlin-Heidelberg, pp. 175–187.

Levitt P Rakic P (1980) Immunoperoxidase localization of glial fibrillary acidic protein in radial glial cells and astrocytes of the developing Rhesus monkey brain. J Comp Neurol 193:815–840.

Liuzzi FJ Miller RH (1987) Radially oriented astrocytes in the adult rat spinal cord. Brain Res 403:385–388.

McDermott KWG Lantos PL (1989) The distribution of GFAP and vimentin in postnatal marmoset (Callithrix jacchus) brain. Dev Brain Res 45:169–177.

Meshul CK Seil FJ Herndon RM (1987) Astrocytes play a role in regulation of synaptic density. Brain Res 402:139–145.

Miller RH Liuzzi FJ (1986) Regional specialization of the radial glial cells of the adult frog spinal cord. J Neurocytol 15:187–196.

Misson JP Edwards MA Yamamoto M Caviness VS (1988) Mitotic cycling of radial glial cells of the fetal murine cerebral wall: a combined autoradiographic and immunohistochemical study. Dev Brain Res 38:183–190.

Monzón-Mayor M Yanes C Ghandour MS De Barry J Gombos G (1990) Glial fibrillary acidic protein and vimentin immunohistochemistry in the developing and adult midbrain of the lizard Gallotia galloti. J Comp Neurol 294:1–11.

Nona SN Shehab SAS Stafford CA Cronly-Dillon JR (1989) Glial fibrillary acidic protein (GFAP) from goldfish: its localization in visual pathway. Glia 2:189–200.

Onteniente B Kimura H Maeda T (1983) Comparative study of the glial fibrillary acidic protein in vertebrates by PAP immunohistochemistry. J Comp Neurol 215:427–436.

Peters A Palay SL Webster HF (1976) The Fine Structure of the Nervous System: Neurons and Supporting Cells. WB Saunders, Philadelphia.

Raine CS (1984) On the association between perinodal astrocytic processes and the node of Ranvier in the CNS. J Neurocytol 13:21–27.

Rakic P (1981) Neuronal-glial interaction during brain development. Trends Neurosci 4:184–187.

Ramón y Cajal S (1909–1911) Histologie du système nerveux de l'homme et des vertébrés. Reimp. CSIC Madrid 1952–1955.

Ramón y Cajal S (1919) Nota sobre las epitelio fibrillas del epéndimo. Trab Lab Inv Biol 17:87–94.

Reichenbach A (1990) Radial glial cells are present in the velum medullare of adult monkeys. J Hirnforsch 31:269–271.

Roessmann U Velasco ME Sindley SD Gambetti P (1980) Glial fibrillary acidic protein (GFAP) in ependymal cells during development. An immunocytochemical study. Brain Res 200:13–21.

Roots B (1982) Comparative studies on glial markers. J Exp Biol 95:167–180.

Roots B (1986) Phylogenetic development of astrocytes, in Astrocytes. Development, Morphology and Regional Specialization of Astrocytes. vol. 1. (Fedoroff S Vernadakis A, eds.), Academic, Orlando, FL, pp. 1–34.

Rubio M Suárez I Bodega G Fernández B (1992) Glial fibrillary acidic protein and vimentin immunohistochemistry in the posterior rhombencephalon of the iberian barb *(Barbus comiza)*. Neurosci Lett 134:203–206.

Sarnat HB (1992) Regional differentiation of the human fetal ependyma: immunocytochemical markers. J Neuropathol Exp Neurol 51:58–75.

Sensenbrenner M Develliers G Bock K Porte A (1980) Biochemical and ultrastructural studies of cultures rat astroglial cells. Effect of brain extract and dibutyryl cyclic AMP on glial fibrillary acidic protein and glial filament. Differentiation 17:51–61.

Stensaas LJ Stensaas SS (1968) Light microscopy of glial cells in turtles and birds. Z Zellforsch 91:315–340.

Suárez I Fernández B (1983) Structure and ultrastructure of the external glial layer in the hypothalamus of the hamster. J Hirnforsch 24:99–109.

Suárez I Raff MC (1989) Subpial and perivascular astrocytes associated with nodes of Ranvier in the rat optic nerve. J Neurocytol 18:577–582.

Suárez I Fernández B Garcia-Segura LM (1980) Specialized contacts of astrocytes with astrocytes and with other cell types in the hypothalamus of the hamster. J Anat 130:55–61.

Suárez I Fernández B Bodega G Tranque P Olmos G García-Segura LM (1987) Postnatal development of glial fibrillary acidic protein immunoreactivity in the hamster arcuate nucleus. Dev Brain Res 37:89–95.

Suárez I Bodega G Arilla E Rubio M Villalba RM Fernández B (1992) Different response of astrocytes and Bergmann glial cells to portacaval shunt: an immunohistochemical study in the rat cerebellum. Glia 6:172–179.

Suárez I Bodega G Rubio M Garcia-Segura LM Fernández B (1994) Astroglial induction of angiogenesis in vivo. J Neural Transp Plast 5:1–10.

Vernadakis A (1986) Changes in astrocytes with aging, in Astrocytes. Development, Morphology and Regional Specialization of Astrocytes. vol 1. (Fedoroff S Vernadakis A, eds.), Academic, Orlando, FL, pp. 377–407.

Voight T (1989) Development of glial cells in the cerebral wall of ferrets: direct tracing of their transformation from radial glia into astrocytes. J Comp Neurol 289:74–88.

Wolburg H Kästner R Kurz-Isler G (1983) Lack of orthogonal particle assemblies and presence of tight junctions in astrocytes of goldfish. A freeze-fracture study. Cell Tissue Res 234:389–402.

Wujek JR Reier PJ (1984) Astrocytic membrane morphology: differences between mammalian and amphibian astrocytes after axotomy. J Comp Neurol 222:607–619.

Yanes C Monzón-Mayor M Ghandour MS De Barry J Gombos G (1990) Radial glia and astrocytes in developing and adult telencephalon of the lizard *Gallotia galloti* as revealed by immunohistochemistry with anti-GFAP and anti-vimentin antibodies. J Comp Neurol 295:559–568.

Zamora AJ Mutin M (1988) Vimentin and glial fibrillary acidic protein filaments in radial glia of the adult urodele spinal cord. Neuroscience 27:279–288.

The Role and Fate of Radial Glial Cells During Development of the Mammalian Cortex

Thomas Voigt and Ana D. de Lima

1. Identification of Radial Glial Cells

The existence of long, radially oriented fibers spanning the entire thickness of the developing fetal cerebrum has been recognized in many vertebrate species since the introduction of the Golgi technique. These specialized glial cells have their cell bodies in the ventricular zone among the dividing neuroblasts, where they are attached via an endfoot to the ventricular surface. With one long process they traverse the entire thickness of the cortical wall and terminate with several endfeet on the pial surface (or occasionally on blood vessels) (Figs. 1 and 2). Throughout the entire period of neurogenesis, these cells can be found within the cortical wall. In some species (e.g., primates and humans) radial glial cells can reach several millimeters in length (Rakic, 1972). Although over time a variety of names have been given to these cells (epithelial cells, spongioblasts, matrix cells, fetal ependymal cells, radial cells) the name radial glial cell is now generally accepted (for a review, *see* Schmechel and Rakic, 1979). In the cerebral cortex of mammals, radial glial cells disappear at the end of neuronal cell migration. In contrast, in some nonmammalian species radial glia may persist into adulthood, expanding radial processes from the ventricular wall to the pial surface (Yanes et al., 1990). Radial glial cells are considered to be the most primi-

From: *Neuron-Glia Interrelations During Phylogeny: I. Phylogeny and Ontogeny of Glial Cells*
A. Vernadakis and B. Roots, Eds. Humana Press Inc., Totowa, NJ

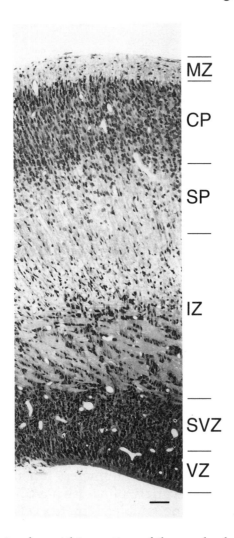

Fig. 1. Nissl-stained semithin section of the cerebral cortex of a new-born ferret. Pial surface is to the top and ventricular surface to the bottom. The marginal zone (MZ) is a thin cell-poor layer beneath the pial surface. At this age the cortical plate (CP) includes in its most superficial part a cell-dense layer that contains the youngest cortical cells. The lowest part of the cortical plate contains cells that will form deep cortical layers. Some authors denominate only the upper part with densely packed cells as cortical plate and designate already differentiating layers by their numbers. MZ = marginal zone; SP = subplate zone; IZ = intermediate zone; SVZ = subventricular zone; VZ = ventricular zone. Scale bar = 50 μm.

tive neuroglial elements in vertebrates and represent the principal glia in the teleost and reptile central nervous system (Onteniente et al., 1983). In this chapter, the authors will focus on the function and development of radial glial cells in the mammalian cerebral cortex, but references to some relevant data of other vertebrate groups will be included.

In the mammalian cerebral cortex radial glial cells possess specific immunocytochemical properties and can be recognized as an entire population. In primates they are labeled by an antibody against glial fibrillary acidic protein (GFAP), an intermediate filament also present in mature astrocytes (Antanitus et al., 1976; Choi and Lapham, 1978; Levitt and Rakic, 1980). In nonprimate mammals, such as mouse, rat, cat, and ferret, radial glial cells do not stain with GFAP antibodies but can be identified using antibodies against the intermediate filament protein vimentin (Fig. 3). In these species GFAP is present in differentiated astrocytes (Bignami and Dahl, 1974a; Dahl, 1981; Dahl et al., 1981; Schnitzer et al., 1981; Bignami et al., 1982; Cochard and Paulin, 1984; Pixley and Vellis, 1984; Engel and Müller, 1989; Stichel et al., 1991). Recently, additional monoclonal antibodies (Rat-401, RC1, RC2) have been developed that label radial glial cells with similar or better specificity than the antivimentin antibodies (Hockfield and McKay, 1985; Misson et al., 1988a,b; Edwards et al., 1990; Misson et al., 1991).

2. Function of Radial Glial Cells

2.1. Structural Support

A typical feature of young brain tissue is its softness, caused by large extracellular spaces between the cells and the absence of myelinated fiber tracts. Although loose packaging of cells is likely to be a prerequisite for the massive cellular and axonal movement taking place during the time of early development, the stability of the structure is also essential for organized growth. A key stabilization factor during the period of tissue expansion is the radially oriented glial elements spanning the whole thickness of the cortical wall (Peters et al., 1991). Immunohistochemical investigations have shown that embryonic mouse radial fibers are spaced at regular intervals of about 8–10 μm. This distance of 1–2

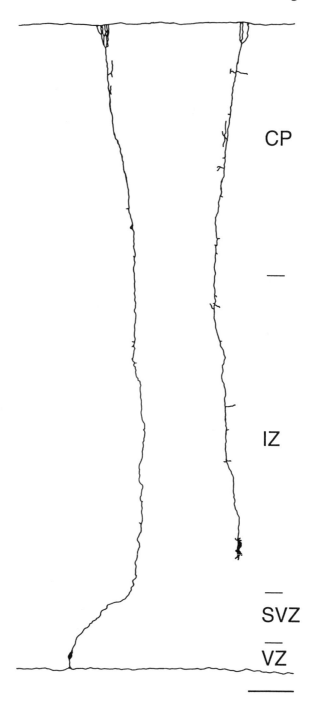

cell diameters is kept fairly constant over the course of development (Gadisseux et al., 1989). The geometry and constant density of radial glial cells is in agreement with the hypothesis that these cells form a scaffold that stabilizes the cortex during its growth.

2.2. Support of Cellular Migration

The organized migration of neuroblasts from the proliferative zone to their final locations at the lower limit of the marginal zone is perhaps one of the most remarkable developmental events during cortical histogenesis. During development, neurons reach the pial side of the cortex in different ways. Initially, when distances are fairly short and all neuroepithelial cells span the entire thickness of the cortical wall, postmitotic nerve cells reach the pial side by translocation of their nucleus (Morest, 1970). As the thickness of the cortical wall increases, neurons lose contact with the pial surface. In order to reach the upper portion of the cortical plate at this later stage of development, neurons attach to radial glial cells and migrate actively along the oriented glial processes (Rakic, 1971, 1972). The guidance of neuronal migration along glial cells has been confirmed in the mammalian cerebral cortex as well as in other structures, both *in situ* (Gadisseux et al., 1990), and in vitro (Hatten et al., 1984, 1986; Edmondson and Hatten, 1987) and has been reviewed extensively (*see,* Sidman and Rakic, 1973; Rakic, 1981, 1990)

2.3. Metabolic Support of Migrating Neurons

Electron microscopic observations of the developing cortex have revealed that radial glia are abundantly filled with glycogen, whereas migratory and postmigratory neurons are devoid of intracytoplasmic glycogen granules (Gadisseux et al., 1989;

Fig. 2. (*previous page*) Drawings of two radial glial cells in different stages of development. These cells were labeled after light paraformaldehyde fixation by putting small DiI crystals on the cortical surface of P2 (postnatal day 2) ferrets. The left cell is still in contact with the ventricular surface, whereas the right cell has lost this contact and the cell body is displaced into the intermediate zone. MZ = marginal zone; CP= cortical plate; IZ = intermediate zone; SVZ = subventricular zone; VZ = ventricular zone. Scale bar = 100 µm.

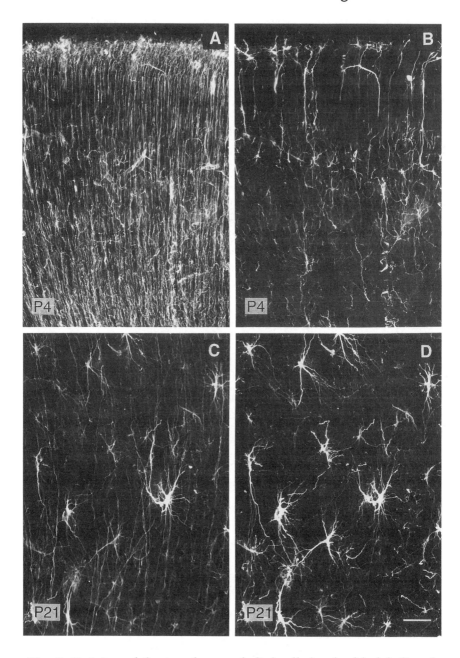

Fig. 3. Staining of the two forms of glial cells by double labeling for vimentin and GFAP using antibodies linked to two different chromophores. At P4 vimentin containing radial glial fibers are still abundant **(A)**, whereas GFAP-immunolabeled cells **(B)** increase dramatically in

Gressens and Evrard, 1993). In contrast to neurons, which apparently have no direct access to energy sources, radial glial fibers contact the vascular system on either side of the cortical wall through their endfeet. In this way, radial glia have access to carbohydrates from the bloodstream, needed to form glycogen. Based on this observation, it was suggested that glial fascicles may supply energy for neurons migrating along their surfaces (Gressens and Evrard, 1993).

2.4. Guidance of Axons

During the period of neuronal migration, extrinsic axons begin to enter the cortex. Within the intermediate zone, axonal growth is mostly orthogonal to the direction of neuronal migration. However, individual axons that leave the main fiber trajectory assume an ascending radial orientation and innervate their target cells in the overlaying cortical plate. Based on results of anterograde transport studies, it has been suggested that radial glial cells may help guiding axons from the intermediate zone into their cortical target area. Electronmicroscopic reconstructions of peroxidase-labeled callosal afferent fibers revealed that the growth cones of these axons attached in the intermediate zone to individual glial fibers and follow them into the cortical plate (Norris and Kalil, 1991). It is likely that this mechanism of guidance is generally used by both ingrowing and outgrowing axons throughout the developing CNS. In rats and hamsters, radial glial cells in the diencephalon and mesencephalon are labeled transcellularly when retinal axons are anterogradely labeled with peroxidase or DiI (Kageyama and Robertson, 1993). Although these experiments do not themselves show that radial glia guide the retinal axons, they show the existence of a close contact between axons and glial cells. The lamellate expansions observed in Golgi and DiI stained material might be involved in this guid-

---

number compared to younger ages (not illustrated). All GFAP-labeled processes are also labeled with vimentin. Vimentin (C) and GFAP (D) immunostaining of the same field in the lower portion of the cortical plate at P21. The density of vimentin labeled radial glial cells is greatly reduced compared to earlier stages, whereas the GFAP/vimentin positive young astrocytes have become more mature and numerous. Scale bar = 25 μm.

ance. These fine extensions of radial glia are oriented horizontally (Rakic, 1972, 1989; Voigt, 1989; Roberts et al., 1993) and are especially numerous in the intermediate zone, where the bulk of axons enter the cortical plate.

2.5. Sculpturing Factor
for Cortical Cytoarchitecture

Anatomical studies of mammals and other vertebrates (Goffinet, 1983, 1984) provide convincing evidence for a role of radial glia in organizing the cortical architecture. Cells in the mammalian cortical plate and cerebral cortex have a predominantly vertical orientation, and it is generally assumed that this orientation is imposed by the orientation of the radial glia that guides neuronal migration (Peters and Sethares, 1991). Indeed, in different reptilian groups the degree of cellular orientation can be correlated subtly with the orientation and ramifications of radial glia guides (Goffinet, 1983, 1984). Additionally, the existence of pyramidal cell modules in which apical dendrites form radially oriented clusters has been related to the mode of migration of groups of neurons along the same radial glial fiber (e.g., Peters and Feldman, 1973; Rakic, 1984; Peters and Sethares, 1991).

The radial organization of the glial fibers imposes not only a strict orientation to cells that migrate along the aligned paths into the cortical plate: It also forms a point-to-point projection between the ventricular germinal zone and the overlying cortical mantle (Rakic, 1971, 1972, 1974). The fact that neuronal migration is constrained by the orientation of radial glial fibers has led to the conjecture that these cells may ensure faithful mapping of the ventricular surface in the expanding and convoluted cerebral cortex (Rakic, 1978). According to this hypothesis, the germinal zone contains a mosaic of proliferative units that form a protomap (Reznikov et al., 1984; Rakic, 1988). In this way the radial arrangement of radial glia would presage the columnar functional organization of the mature cortex, and a hypothetical primordial topography could be transposed in a one-to-one relationship from the ventricular zone to the cortical plate. The original protomap hypothesis required that clonally related cells migrate preferentially along single radial glial cells (Rakic, 1988). Recent experiments with tagged retroviruses, as well as slice experiments using video microscopy, have shown that this is not strictly the case. Cells originating from the same stem cell can end up in cortical positions far apart from each other (Walsh and Cepko, 1988; O'Rourke et al., 1992; Walsh and Cepko, 1992; Fishell et al., 1993; Roberts et al., 1993; Walsh and Cepko, 1993). Thus, horizontal and oblique migration of some neurons may lead to substantial dispersion of clonally related neurons.

Although the strict interpretation (Rakic, 1978) of the protomap or radial unit hypothesis does not hold, recent studies (Arimatsu et al., 1992; Ferri and Levitt, 1993) support a revised version of this hypothesis (Rakic, 1988, 1992) showing an early regional specification of neuronal precursors in the telencephalic wall in the absence of extrinsic stimuli (see also Price, 1993).

3. Development of Radial Glial Cells

During development the cortical plate starts as a thin neuroepithelium and, depending on the species, grows to a thickness of up to several millimeters. During the entire period radial fibers bridge the distance from the ventricular zone to the pial surface with little variation in the spacing between fibers (Gadisseux et al., 1989). To guarantee constant spacing within the scaffold, the population of radial glial cells must adjust to both the vertical and the lateral expansion of the cortical wall. Apparently several mechanisms have been adopted to cope with this problem. During the early period of neurogenesis radial glial fibers are slightly coiled and run in fascicles (Fig. 4A) (Gadisseux et al., 1989). When the cortical wall thickens the coiled fibers stretch like springs. Lateral expansion is compensated for by defasciculation of the fibers (Fig. 4B) (Gadisseux et al., 1989). Beginning with the second half of gestation, fibers of the radial glial cells can no longer follow the increase in thickness. At this time they lose their contacts with the ventricular surface and start to transform into astrocytes (Figs. 2 and 4C) *(see below)*. To compensate for this loss of fibers, new radial glial cells are generated within the ventricular zone (Levitt et al., 1981, 1983; Misson et al., 1988a). The high percentage of dividing glial cells found at midgestation of the primate cortex (80% of dividing cells) indicates that substantial replacement takes place during this time (Levitt et al., 1983). Once born, the young fibers grow orienting along older fibers toward

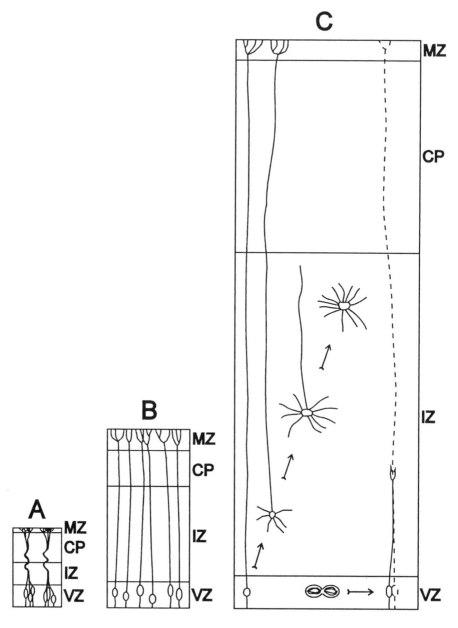

Fig. 4. Summary diagram showing the development of radial glial cells in the context of cortical development. Early during cortical development radial glial cells are curled like springs and are arranged in fasciculated bundles **(A)**. Both defasciculation and stretching compensates for the vertical and horizontal expansion of the cortex, guar-

the pial surface (Takahashi et al., 1990). Thus, by the dynamic processes of transformation and regrowth, a constant number of fibers per volume of cortex is maintained during the endphase of neurogenesis, when the expansion of this structure is the greatest. The regrowth of new radial glial cells explains the otherwise puzzling fact that at later developmental stages radial fibers often tend to end at blood vessels located somewhere in the middle of the cortical wall (Schmechel and Rakic, 1979). During their elongation, radial glial fibers occasionally hit a blood vessel and attach to it instead of continuing their growth to the more distant pial surface.

4. The Fate of Radial Glial Cells

Immunohistochemical as well as Golgi data indicate that two major events in glial development take place during the formation of the cerebral cortex. Although radial glial cells are abundant during the period of cellular migration, their number continuously declines once cell proliferation comes to an end (Fig. 3). When all cells have reached the cortical plate, radial glial fibers disappear entirely. Coincident with the decline of the radial glia, young astrocytes start to appear within the cortical wall (Choi and Lapham, 1978; Schmechel and Rakic, 1979; Levitt and Rakic, 1980; Misson et al., 1988b; Voigt, 1989; Takahashi et al., 1990; Stichel et al., 1991).

By doing electron microscopic analyses of tissue labeled with the anti-GFAP antibody, Choi and Lapham (1978) found that radial glial cells in humans contain some characteristic features of

---

anteeing evenly spaced radial glial fibers that span the entire thickness of the cortical wall (**B**). With further expansion of the cortical wall the compensation by stretching and defasciculation of radial glial fibers becomes insufficient. Radial glial fibers begin to lose their contact with the outer surfaces and transform into astrocytes (**C**, left side). To compensate for the loss of fibers new radial glial cells are generated (**C**, right side). Their elongation is guided by existing radial glial cells until they reach a blood vessel or the pial surface. At the end of histogenesis the even spacing of radial glial cells is achieved by a constant turnover of the entire population. MZ = marginal zone; CP = cortical plate; IZ = intermediate zone; VZ = ventricular zone.

astrocytes. During the period when radial glial cells decrease in number, many cells have an appearance intermediate between those of radial glia and astrocytes. These observations led Choi and Lapham to formulate the hypothesis that radial glia do not disappear during development, but instead gradually transform into astroglial cells. Comparable observations have been made in primates (Schmechel and Rakic, 1979; Levitt and Rakic, 1980), rats (Pixley and Vellis, 1984; Stichel et al., 1991), cats (Engel and Müller, 1989), and ferrets (Voigt, 1989).

In rodents and carnivores, the change in glial morphology is linked to a change in the intermediate filament protein expressed by the cell (Fig. 3). Double label experiments with antibodies directed against the intermediate filaments vimentin and GFAP have shown that in nonprimates radial glial cells contain only vimentin, whereas mature astrocytes contain only the intermediate filament GFAP (Bignami and Dahl, 1974a, 1974b; Dahl, 1981; Dahl et al., 1981; Schnitzer et al., 1981; Cochard and Paulin, 1984; Pixley and Vellis, 1984; Voigt, 1989; Stichel et al., 1991). Young astrocytes, that appear at the end of cell proliferation, contain both forms of intermediate filaments (Voigt, 1989), further supporting the transformation hypothesis. Although in rats the colocalization of vimentin and GFAP in young astrocytes seems to be less prominent than in the ferret (Pixley and Vellis, 1984), in both species a switch in the expression of intermediate filament protein occurs at a similar time in glial development. At present it is not known whether primate radial glial cells express only GFAP or whether they also express vimentin.

Although the sequence of events described using immuno-histochemical methods does support the idea that radial glial cells transform into astrocytes, alternative explanations are not refuted by these results. Vimentin positive radial glial cells could, for instance, be eliminated by cell death, whereas young astrocytes containing vimentin and GFAP are generated from precursor cells.

Application of tiny DiI crystals to the pia of fixed ferret brains frequently labels radial glial fibers (Fig. 2). The dye is taken up by a cell via its endfeet and labels the entire fiber in a Golgi-like fashion. Shortly after birth most of the radial glial cells have their cell bodies within the ventricular zone (Fig. 2). At later developmental stages many cells have retracted somata as if in an early stage of cell transformation (Fig. 2). These observations lead us to the idea that it should be possible to test the transformation hypothesis more directly by combining immunohistochemical and tracing techniques. If radial glial cells transform into astrocytes, a dye that has been incorporated into the cell bodies of radial glial cells should be present in astrocytes after the transformation. We performed these experiments in ferrets. These animals are ideally suited for this kind of study since they are born very immature compared to most other species (Greiner and Weidman, 1981; Linden et al., 1981). At the time of birth the ferret cortex contains only radial glial cells, whereas 3 weeks after birth most radial glial cells have disappeared and astrocytes are the predominant form of glia (Voigt, 1989). The application of tracing dyes, such as Fluoro-gold, Fast blue, and Rhodamine coupled microspheres to the pial surface of a newborn ferret cortex results in the labeling of radial glial cell bodies within the ventricular zone. As is the case with the DiI crystals, the dyes are taken up by the endfeet of the radial glial cell and are transported to the cell body located within the ventricular zone. One to three weeks after application of the dye, counterstaining of brains that have been injected on the day of birth shows that the label is present in newly formed GFAP positive astrocytes (Fig. 5) (Voigt, 1989). The distance from the pia to the location of the double labeled astrocytes was beyond the diffusion range of the dyes. These data strongly support the transformation hypothesis by showing that at least some radial glia cells change morphology to become astrocytes in the adult neocortex.

Recent evidence has been presented for a correlation between the morphological transformation of radial glia and the loss of other radial glial-specific antigens (Misson et al., 1988b; Edwards et al., 1990). This correlation has been confirmed for one antigen (RC1) in culture experiments (Culican et al., 1990). Dissociated cell cultures prepared from mouse cortices on embryonic day 13 contain many radial glial cells, as identified by an antibody specific for radial glia (RC1). These immature cells do not stain for GFAP. After several days in vitro radial glial cells begin to develop branched processes and subsequently acquire the morphology of astrocytes. As in the intact animal this process of transformation is accompanied by a change in intermediate filament (Culican et al., 1990). Interestingly, the transformation from radial glia to

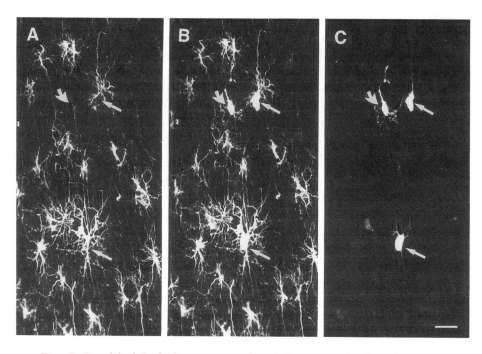

Fig. 5. Double labeled astrocytes (straight arrows) after their trans-formation from radial glial cells. **(A)** GFAP+ astrocytes labeled with a rhodamine linked antibody 3 weeks after Fluoro-gold was injected at P0 under the pial surface. The cells were located in the lower portion of the intermediate zone. **(B)** Double exposure for rhodamine (A) and Fluoro-gold (B) showing the double labeled cells (straight arrows). The cells incorporated the dye via their endfeet at the pial surface when they were radial glial cells. **(C)** Fluoro-gold labeled astrocytes visual-ized with violet UV light only. Notice that only some of the astrocytes are double labeled and that some Fluoro-gold labeled cells with glial morphology do not contain GFAP (curved arrows). Scale bar = 20 μm.

astrocytes in vitro requires the presence of neurons. The role of neurons on the development of glial cells has been intensively studied in the cerebellar cortex (Hatten, 1985; Hatten and Mason, 1986; Gasser and Hatten, 1990; Hatten, 1990).

It is still unclear if the entire population of astrocytes within the adult cortex is generated from radial glial cells, or if prolifera-tion of glial precursor cells within the ventricular zone can gener-ate astrocytes directly. There may well be astrocyte populations that are generated before or after birth directly from immature cells of ventricular or subventricular origin (Goldman and Vaysse, 1991). Additionally, since glial cells generally retain mitogenic capabilities, a differentiated astrocyte may undergo further cell divisions after achieving its final position.

5. Summary and Conclusions

Radial glia are a phylogenetically primitive and ontogenetically immature form of glia. In the cerebral cortex of mammals the radial glia provide a structural framework into which the cellular elements can develop. They confer stability for the entire structure and serve to guide neurons and axons on their way to their final positions. By their vertical orientation radial glia help to impose a predominantly vertical orientation to neocortical cells, a likely prerequisite for the proper formation of synaptic connections.

The majority of radial glial cells are exclusively vimentin-positive when they span the cortical wall. During the last third of cortical histogenesis, some radial glial cell bodies move away from the ventricular zone, an event that may mark the beginning of the morphological transformation into astrocytes. During this transformation, the long tube-like radial glial cells become spherical multipolar cells with processes that extend for only a short distance from the cell body. At this time these cells start to express both the astrocyte marker GFAP and the radial glial marker vimentin. As they mature vimentin is completely substituted by GFAP, which remains as the major intermediate filament protein throughout life. Additional changes in antigen expression have recently been found to be associated with the morphological transformation of radial glial cells in cortex.

Although the causal relationship between morphological transformation and changes in certain antigen expression remains unclear, it is interesting that morphological adaptations also appear coincident with changes in the relationship of astrocytes to neurons. The expansion of the nonradial network of astroglial fibers through the cortical layers in the perinatal period coincides with the rapid acceleration of growth and differentiation of cortical neurons and occurs simultaneously with the formation of intracortical circuitry. Thus, the signals that trigger these radical transformations in the astroglial population may emanate from the developing neurons of the cortical strata.

References

Antanitus DS Choi BH Lapham LW (1976) The demonstration of glial fibrillary acidic protein in the cerebrum of the human fetus by indirect immunofluorescence. Brain Res 103:613–616.

Arimatsu Y Miyamoto M Nihonmatsu I Hirata K Uratani Y Hatanaka Y Takiguchi-Hayashi K (1992) Early regional specification for a molecular neuronal phenotype in the rat neocortex. Proc Natl Acad Sci USA 89:8879–8883.

Bignami A Raju T Dahl D (1982) Localization of vimentin, the nonspecific intermediate filament protein, in embryonal glia and in early differentiating neurons. Dev Biol 91:286–295.

Bignami A Dahl D (1974a) Astrocyte-specific protein and neuroglial differentiation. An immunofluorescence study with antibodies to the glial fibrillary acidic protein. J Comp Neurol 153:27–38.

Bignami A Dahl D (1974b) Astrocyte-specific protein and radial glia in the cerebral cortex of newborn rat. Nature 252:55–56.

Choi BH Lapham LW (1978) Radial glia in the human fetal cerebrum: A combined golgi, immunofluorescent and electron microscopic study. Brain Res 148:295–311.

Cochard P Paulin D (1984) Initial expression of neurofilaments and vimentin in the central and peripheral nervous system of the mouse embryo in vivo. J Neurosci 4:2080–2094.

Culican SM Baumrind NL Yamamoto M Pearlman AL (1990) Cortical radial glia: Identification in tissue culture and evidence for their transformation to astrocytes. J Neurosci 10:684–692.

Dahl D (1981) The vimentin-GFA protein transition in rat neuroglia cytoskeleton occurs at the time of myelination. J Neurosci Res 6: 741–748.

Dahl D Rueger DC Bignami A Weber K Osborn M (1981) Vimentin, the 57000 molecular weight protein of fibroblast filaments, is the major cytoskeletal component in immature glia. Eur J Cell Bio 24:191–196.

Edmondson JC Hatten ME (1987) Glial-guided granule neuron migration in vitro: a high-resolution time-lapse video microscopic study. J Neurosci 7:1928–1934.

Edwards MA Yamamoto M Caviness VS Jr. (1990) Organization of radial glia and related cells in the developing murine CNS. An analysis based upon a new monoclonal antibody marker. Neurosci 36:121–144.

Engel AK Müller CM (1989) Postnatal development of vimentinimmunoreactive radial glial cells in the primary visual cortex of the cat. J Neurocytol 18:437–450.

Ferri RT Levitt P (1993) Cerebral cortical progenitors are fated to produce region-specific neuronal populations. Cereb Cortex 3:187–198.

Fishell G Mason CA Hatten ME (1993) Dispersion of neural progenitors within the germinal zones of the forebrain. Nature 362:636–638.

Gadisseux J-F Evrard P Misson J-P Caviness VS Jr (1989) Dynamic structure of the radial glial fiber system of the developing murine cerebral wall. An immunocytochemical analysis. Dev Brain Res 50:55–67.

Gadisseux J-F Kadhim HJ Van den Bosch de Aguilar P Caviness VS Jr Evrard P (1990) Neuron migration within the radial glial fiber system of the developing murine cerebrum: An electron microscopic autoradiographic analysis. Dev Brain Res 52:39–56.

Gasser UE Hatten ME (1990) Neuron-glia interactions of rat hippocampal cells *in vitro*: Glial-guided neuronal migration and neuronal regulation of glial differentiation. J Neurosci 10:1276–1285.

Goffinet AM (1983) The embryonic development of the cortical plate in reptiles: a comparative study in *Emys orbicularis* and *Lacerta agilis*. J Comp Neurol 215:437–452.

Goffinet AM (1984) The embryonic development of the cerebral cortex: what we can learn from reptiles? in Organizing Principles of Neural Development (Sharma SC, ed.), Plenum, New York, pp. 251–259.

Goldman JE Vaysse PJ-J (1991) Tracing glial cell lineages in the mammalian forebrain. Glia 4:149–156.

Greiner JV Weidman TA (1981) Histogenesis of ferret retina. Exp Eye Res 33:315–332.

Gressens P Evrard P (1993) The glial fascicle: an ontogenic and phylogenic unit guiding, supplying and distribution mammalian cortical neurons. Dev Brain Res 76:272–277.

Hatten ME Liem RKH Mason CA (1984) Two forms of cerebellar glial cells interact differently with neurons in vitro. J Cell Biol 98:193–204.

Hatten ME (1985) Neuronal regulation of astroglial morphology and proliferation in vitro. J Cell Biol 100:384–396.

Hatten ME Liem RKH Mason CA (1986) Weaver mouse cerebellar granule neurons fail to migrate on wild-type astroglial processes in vitro. J Neurosci 6:2676–2683.

Hatten ME (1990) Riding the glial monorail: A common mechanism for glial-guided neuronal migration in different regions of the developing mammalian brain. Trends Neurosci 13:179–184.

Hatten ME Mason CA (1986) Neuron-astroglia interactions *in vitro* and *in vivo*. Trends Neurosci 9:168–174.

Hockfield S McKay RDG (1985) Identification of major cell classes in the developing mammalian nervous system. J Neurosci 5:3310–3328.

Kageyama GH Robertson RT (1993) Transcellular retrograde labeling of radial glial cells with WGA-HRP and DiI in neonatal rat and hamster. Glia 9:70–81.

Levitt P Cooper ML Rakic P (1981) Coexistence of neuronal and glial precursor cells in the cerebral ventricular zone of the fetal monkey: An utlrastructural immunoperoxidase analysis. J Neurosci 1:27–39.

Levitt P Cooper ML Rakic P (1983) Early divergence and changing proportions of neuronal and glial precursor cells in the primate cerebral ventricular zone. Dev Biol 96:472–484.

Levitt P Rakic P (1980) Immunoperoxidase localization of glial fibrillary acidic protein in radial glial cells and astrocytes of the developing rhesus monkey brain. J Comp Neurol 193:815–840.

Linden DC Guillery RW Cucchiaro J (1981) The dorsal lateral geniculate nucleus of the normal ferret and its postnatal development. J Comp Neurol 203:189–211.

Misson J-P Edwards MA Yamamoto M Caviness VS Jr (1988a) Mitotic cycling of radial glial cells of the fetal murine cerebral wall: a combined autoradiographic and immunohistochemical study. Dev Brain Res 38:183–190.

Misson J-P Edwards MA Yamamoto M Caviness VS Jr. (1988b) Identification of radial glial cells within the developing murine central nervous system: studies based upon a new immunohistochemical marker. Dev Brain Res 44:95–108.

Misson J-P Takahashi T Caviness VS Jr (1991) Ontogeny of radial and other astroglial cells in murine cerebral cortex. Glia 4:138–148.

Morest DK (1970) A study of neurogenesis in the forebrain of opossum pouch young. Z Anat Entwickl -Gesch 130:265–305.

Norris CR Kalil K (1991) Guidance of callosal axons by radial glia in the developing cerebral cortex. J Neurosci 11:3481–3492.

O'Rourke NA Dailey ME Smith SJ McConnell SK (1992) Diverse migratory pathways in the developing cerebral cortex. Science 258:299–302.

Onteniente B Kimura H Maeda T (1983) Comparative study of the glial fibrillary acidic protein in vertebrates by PAP immunohistochemistry. J Comp Neurol 215:427–436.

Peters A Palay SL Webster HdeF (1991) The Fine Structure of the Nervous System. Neurons and Their Supporting Cells, Oxford University Press, New York, pp. 273–311.

Peters A Feldman M (1973) The cortical plate and molecular layer of the late rat fetus. Z Anat Entwickl -Gesch 141:3-37.

Peters A Sethares C (1991) Organization of pyramidal neurons in area 17 of monkey visual cortex. J Comp Neurol 306:1–23.

Pixley SKP Vellis J (1984) Transition between immature radial glia and mature astrocytes studied with a monoclonal antibody to vimentin. Dev Brain Res 15:201–209.

Price J (1993) Organizing the cerebrum. Nature 362:590–591.

Rakic P (1971) Guidance of neurons migrating to the fetal monkey neocortex. Brain Res 33:471–476.

Rakic P (1972) Mode of cell migration to the superficial layers of fetal monkey neocortex. J Comp Neurol 145:61–84.

Rakic P (1974) Neurons in rhesus monkey visual cortex: Systematic relation between time of origin and eventual disposition. Science 183:425–427.

Rakic P (1978) Neuronal migration and contact guidance in the primate telencephalon. Postgraduate Medical Journal 54:25–40.

Rakic P (1981) Neuronal-glia interaction during brain development. Trends Neurosci 4:184–187.

Rakic P (1984) Organizing principles for development of primate cerebral cortex, in Organizing Principles of Neural Development (Sharma SC, ed.), Plenum, New York, pp. 21–48.

Rakic P (1988) Specification of cerebral cortical areas. Science 241:170–176.

Rakic P (1989) Emergence of neuronal and glial cell lineages in primate brain, in Cellular and Molecular Biology of Neuronal Development (Black IB, ed.), Plenum, New York, pp. 29–50.

Rakic P (1990) Principles of neural cell migration. Experientia 46:882–891.

Rakic P (1992) Dividing up the neocortex. Science 258:1421,1422.

Reznikov KY Fülöp Z Hajós F (1984) Mosaicism of the ventricular layer as the developmental basis of neocortical columnar organization.l. Anat Embryol (Berl) 170:99–105.

Roberts JS O'Rourke NA McConnell SK (1993) Cell migration in cultured cerebral cortical slices. Dev Biol 155:396–408.

Schmechel DE Rakic P (1979) A Golgi study of radial glial cells in developing monkey telencephalon: Morphogenesis and transformation into astrocytes. Anat Embryol (Berl) 156:115–152.

Schnitzer J Franke WW Schachner M (1981) Immunocytochemical demonstration of vimentin in astrocytes and ependymal cells of developing and adult mouse nervous system. J Cell Biol 90:435–447.

Sidman RL Rakic P (1973) Neuronal migration, with special reference to developing human brain: a review. Brain Res 62:1–35.

Stichel CC Müller CM Zilles K (1991) Distribution of glial fibrillary acidic protein and vimentin immunoreactivity during rat visual cortex development. J Neurocytol 20:97–108.

Takahashi T Misson J-P Caviness VS Jr (1990) Glial process elongation and branching in the developing murine neocortex: A qualitative and quantitative immunohistochemical analysis. J Comp Neurol 302:15–28.

Voigt T (1989) Development of glial cells in the cerebral wall of ferrets: Direct tracing of their transformation from radial glia into astrocytes. J Comp Neurol 289:74–88.

Walsh C Cepko CL (1988) Clonally related cortical cells show several migration patterns. Science 241:1342–1345.

Walsh C Cepko CL (1992) Widespread dispersion of neuronal clones across functional regions of the cerebral cortex. Science 255:434–440.

Walsh C Cepko CL (1993) Clonal dispersion in proliferative layers of developing cerebral cortex. Nature 362:632–635.

Yanes C Monzon-Mayor M Ghandour MS De Barry J Gombos G (1990) Radial glia and astrocytes in developing and adult telencephalon of the lizard *Gallotia galloti* as revealed by immunohistochemistry with anti-GFAP and anti-vimentin antibodies. J Comp Neurol 295:559–568.

Astrocyte Differentiation and Correlated Neuronal Changes in the Opossum Superior Colliculus

Leny A. Cavalcante
and Penha C. Barradas

1. Introduction

The life history of opossums (genus *Didelphis*) provides a unique opportunity for developmental studies on glial cells and on correlated changes in glia and neurons. Marsupials are born after a short period of gestation (13 d in *Didelphis*, McCrady, 1938), with displacement to the postnatal period of early events of ontogenesis, such as the neurogenesis of visual and auditory subcortical nuclei (Cavalcante et al., 1984; Sanderson and Aitkin, 1990). In addition to this time shift, postnatal development of the marsupial central nervous system (CNS) follows a protracted course (Cavalcante, 1987), thus, allowing excellent resolution of developmental sequences.

In addition to the advantages offered by marsupials in developmental studies, they comprise an interesting taxon for study of neurophylogeny. A phylogenetic analysis based on cladistic methodology indicates that marsupials and placentals are sister groups, with the former diverging sooner than the latter from a Late Jurassic/Early Cretaceous common ancestor (Marshall, 1979). That raises the question of whether complete differentiation of astrocytes was achieved in marsupials, particularly in the case of generalized forms such as opossums, or whether they retained reptilian traits (Achucarro, 1915; Roots, 1986).

From: *Neuron-Glia Interrelations During Phylogeny: I. Phylogeny and Ontogeny of Glial Cells*
A. Vernadakis and B. Roots, Eds. Humana Press Inc., Totowa, NJ

An unambiguous answer to the foregoing question may not be possible since astrocytes may be regionally heterogeneous in several respects. This issue has received considerable attention in recent years in what concerns the carrying of membrane receptors for neurotransmitters and neuromodulators (reviewed in Amundson et al., 1992) or molecules of unknown function (Raff, 1989) or the properties as substrata for neuronal growth (reviewed in Qian et al., 1992). However, early work based on metal-impregnated preparations had already unveiled three sources of morphological heterogeneity in (astro)glia:

1. Taxon;
2. Stage of maturation; and
3. Brain region (Ramon y Cajal, 1911; Achucarro, 1915).

In this review, we shall consider aspects of differentiation of the cells, including the heterogeneity of radial glia, in the retino-recipient midbrain tectum (superior colliculus, SC) of the opossum *Didelphis marsupialis* (Cavalcante et al. 1985; Barradas et al., 1988; 1989; Cavalcante and Santos-Silva, 1990; Cavalcante et al., 1991). The retino-recipient midbrain tectum is a favorite model for studies of neuronal development and plasticity in all vertebrates (Stein, 1984; Mendez-Otero et al., 1985; Debski et al., 1990; Simon and O'Leary, 1990). However, it has not been as widely used for studies of glial cell differentiation (but see, for instance, Linser and Perkins, 1987, avian tectum; Monzón-Mayor et al., 1990, reptilian tectum).

To provide a framework for a review of our glial studies, we shall present some preliminaries on the architecture of the retino-recipient midbrain tectum and on development in the opossum superior colliculus (SC).

2. Architecture of the Retino-Recipient Midbrain Tectum

The retino-recipient midbrain tectum comprises the superior colliculus SC of therian mammals—eutherians or placentals and metatherians or marsupials—and the optic tectum of non-mammalian vertebrates plus monotreme or prototherian mammals. These structures, SC and optic tectum, share the feature of being formed by alternate cell-rich and fibrous layers, with this pattern being far more conspicuous in the optic tectum than in its therian counterpart (Ramon y Cajal, 1911; Huber and Crosby, 1943; Oswaldo-Cruz and Rocha-Miranda, 1968; Meek, 1983; Huerta and Harting, 1984).

One of the most flagrant dissimilarities between the SC and the optic tectum is the trajectory of the optic fibers within the midbrain tectum itself. Although the retino-recipient sites are superficial layers in both structures, optic axons course in a layer separated from the pia by a narrow marginal layer in the optic tectum and in the third layer removed from the pia in the SC (Huber and Crosby, 1943; Oswaldo-Cruz and Rocha-Miranda, 1968; Campbell and Hayhow, 1972). This myelo-architectural variance had been attributed to much more vigorous projections from the visual telencephalon in therian mammals than in other vertebrates (Huber and Crosby, 1943) but that may not be applicable to the avian tectum.

As far as mammalian systematics is concerned, the intratectal path of retinal fibers in marsupials reinforces the conclusions derived from cladistic analyses that deny any special relationship with monotremes, that follow the nonmammalian pattern, and place placentals and marsupials as sister groups (Marshall, 1979).

Another difference between the optic tectum and SC lies in the vertical extent of dendritic fields. Such fields may cross several *strata* in the optic tectum (Ramon y Cajal, 1911; Repérant et al., 1981; Meek; 1983) whereas the available evidence in the placental mammals so far studied (Ramon y Cajal, 1911; Huerta and Harting, 1984) and in the opossum (Cavalcante and Bernardes, 1978) shows that dendritic trees in the therian SC fail to display such an across-layers extent.

In spite of the general differences referred to above, the retino-recipient layers proper of SC and optic tectum have common features such as the preponderance of small-sized neurons (Cavalcante and Bernardes, 1978; Repérant et al., 1981; Meek, 1983; Huerta and Harting, 1984), which are probably part of an extensive internuncial network. Retinal terminals comprise from about one-fifth to one-third of all vesicle-containing profiles in the superficial layers in several vertebrates (Repérant et al., 1981 and references; Murray and Edwards, 1982) but have been estimated as representing up to 45% of all terminals in a prosimian (Tigges et al., 1973). Of particular interest is the fact that in the midbrain tectum retinal terminals, together with other types of terminals and their postsynaptic elements, form "synaptic islets" that are not enclosed by glial processes (Behan, 1981; Repérant et al., 1981; Meek, 1983; Schoenitzer and Hollander, 1984; Correa, 1994) in contrast with typical glomeruli in the dorsal lateral geniculate nucleus (Guillery, 1969). However, the opossum SC retinoceptive layers, as their homologue in other mammals, include typical astrocytes, very often adjoining a satellite oligodendrocyte (Fig. 1).

As a corollary to microscopic and ultrastructural similarities, it has been shown that the visuotopic organization of retino-recipient layers is in register with the somatotopic representation of deep layers in both the mammalian SC and the reptilian optic tectum, suggesting a conservative plan of sensory representation in the midbrain (Stein, 1984).

3. General Development of the Superior Colliculus (SC) in the Opossum

The earliest event in the differentiation of central neurons— withdrawal of their precursors from the mitotic cycle—starts in the opossum SC only after birth (Cavalcante et al., 1984), which occurs at the thirteenth postconceptional day (PCD 13). The postnatal neurogenesis of the retino-recipient midbrain tectum is perhaps common to many, if not all, marsupials (brush-tailed possum, Sanderson and Aitkin, 1990), whereas such a timing is unknown among the placentals so far studied (De Long and Sidman, 1962; Cooper and Rakic, 1981; Mustari et al., 1979; Crossland and Umchwat, 1982). An instructive way to look at such taxon-linked variations is to relate the onset and duration of neurogenesis and other developmental events to the cecal period (Robinson and Dreher, 1990), i.e., the time from conception to eye opening (about 80 d). Thus, neurogenesis of the opossum SC starts at about one-sixth of the way along the cecal period, thus, relatively earlier than in the monkey (one-fourth, Cooper and Rakic, 1981), and encompasses more than one eighth of the duration of this period.

Neurogenesis in the opossum SC follows the usual deep-to-superficial gradient found in other mammals, including marsupials (Cavalcante et al., 1984; Sanderson and Aitkin; 1990 and

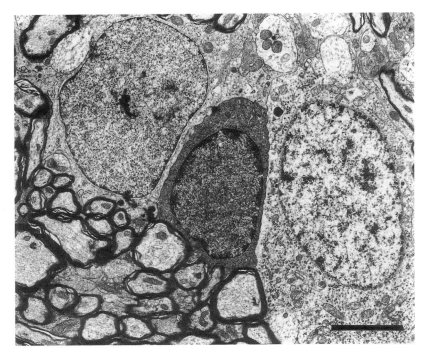

Fig. 1. Astrocyte, oligodendrocyte and neuron at the limit between the superficial gray layer and optic layer (lower left) of the superior colliculus (SC) in an adult opossum. Bar, 2 μm. (From Cavalcante et al., 1991; courtesy of Springer.)

references therein). In addition to disclosing a deep-to-superficial gradient, the relatively long duration of collicular neurogenesis in the opossum has allowed the clarification of a previously conflicting issue, that of tangential gradients (cf, Mustari et al., 1979; Cooper and Rakic, 1981). It has been shown that there is indeed a weak rostrolateral-to-caudomedial gradient of neurogenesis that is restricted to the superficial layers (Cavalcante et al., 1984).

What might be the implications of the features of neurogenesis of the opossum SC? First, confinement of neuron generation (and migration) to the developmental period eliminates the need to maintain arrangements and/or mechanisms for guidance of cellular directional movements during adult life. Such arrangements and/or mechanisms are presumably required for neuronal migration in the optic tectum of adult fishes (Raymond, 1986) and, probably, of amphibians and reptiles. However, the weak tangential gradients of neurogenesis in the opossum SC are hardly sufficient to allow temporal disjunction of neuronal migration to the superficial layers and the arrival of afferent axons (Cavalcante and Rocha-Miranda, 1978a; Mendez-Otero et al., 1985). In fact, at a time of significant neuron production for the superficial layers, by postnatal d 10 (PCD 23), the first contralateral optic fibers are deployed along the entire rostro-caudal and latero-medial extents of the opossum SC.

The ensemble of data from tracing studies of the developing retino-collicular projections, from studies of neurogenesis of retinal ganglion cells and of axonal counts in the developing optic nerve of *Didelphis* (Cavalcante and Rocha-Miranda, 1978a; Mendez-Otero et al., 1985; Kirby et al., 1988; Allodi et al., 1992) have supplied evidence for stages in the development of the optic pathway and retino-collicular projections as follows:

1. Preafferentation stage—slow increase in the number of axons in the optic nerve (before postnatal d 10 [PCD23]);
2. Exclusive innervation from the contralateral retina—moderate increase in the number of optic axons (by postnatal d 10–15 [PCD23 to PCD28]);
3. Exuberant growth and territorial overlap of crossed and uncrossed fibers—abrupt increase in the number of optic axons (by postnatal d 17–23 [PCD30 to PCD36]);
4. Onset and fast progression of arborizations, onset of the restriction of uncrossed fibers to the prospective direct binocular region—maximal numbers of optic axons (by postnatal d 23–30 [PCD36 to PCD43]);
5. Progression of the tangential restriction of uncrossed fibers and binocular radial segregation—fast loss of optic axons (by postnatal d 30–42 [PCD43 to PCD55);
6. Refinement of the patterns of retino-collicular projections—slow loss of optic axons (by postnatal d 42–67 [PCD55 to PCD80], i.e., eye opening).

In addition to the data referred to above, cytoarchitectural differentiation of the SC follows a deep-to-superficial gradient, with separation of the central gray from the SC proper by postnatal d 5 (Mendez-Otero et al., 1985), transient trilamination by day 10 and definition of the optic layer by postnatal d 23 (Cavalcante and Rocha-Miranda, 1978b). Recent ultrastructural work (Correa, 1994) has revealed that terminals with pale mitochondria (R: retinal) are recognized for the first time in postnatal d 30 (PCD43), but that the distinction between R and C (cortical) terminals is not unequivocal before postnatal day 40 (PCD53), thus, at the establishment of binocular segregation (Cavalcante and Rocha-Miranda, 1978a) and concomitant end of the phase of fast loss of optic axons (Kirby et al., 1988).

4. A Dual System of Radial Fibers in the Developing SC, Median Ventricular Formation (MVF), and Lateral or Main Radial System (MRS)

4.1. Differential Uptake of (Optic) Axon-Borne HRP by Radial Glia

Radial glia of the adult amphibian optic tectum may accumulate HRP following an intraocular injection and in the apparent absence of damage to optic axons (Wilczynski and Zakon, 1982). In developing mammals, HRP or WGA-HRP labeling of radial glia-like cells both in the vicinity of the optic chiasm and in retinal target nuclei had often been reported (e.g., Godement et al., 1984; Cavalcante et al., 1985) but not as systematically studied. This apparent neglect may have been owing to difficulties in ruling out extra-axonal routes that might provide access of the label to the capacious extracellular space in the developing brain, or mere damage to the diminutive eye of developing mammals. More important, the concomitant and more conspicuous accumulation of significant amounts of exogenous proteins such as HRP in brain macrophages after its administration by several routes (intracerebral: Valentino and Jones, 1982; Innocenti et al., 1983; bloodstream: Lent et al., 1985) has led to the misguided notion that significant endocytosis is the apanage of macrophages and should be equated with cell death and/or fiber elimination.

Labeling of radial glia in the SC and other retino-recipient sites following HRP injection of the eye of pouch young opossums shows special features and properties (Cavalcante et al., 1985; Cavalcante, 1987; Barradas et al., 1988; 1989). First, cell filling displays regional selectivity, being absent from the collicular midline and adjoining sites (MVF, median ventricular formation, cf, Raedler et al., 1981), irrespective of the density of label in terminal fields adjoining the distal segments and endfeet of radial glia (Fig. 2). Second, away from the midline, cell filling appears in register with terminal fields and shows a direct, although rough, correlation with laterality of the injected eye (Barradas et al., 1989). That helps to define a lateral, main radial system (MRS). Third, cell filling tends to be somatofugal, with clearer delineation of distal processes at longer survival times (cf, Fig. 2 with Fig. 10 in Cavalcante, 1987). Finally, MRS cell filling lasts throughout stages 1–5 of retino-collicular hodogenesis, overlapping the appearance of the first differentiated astrocytes. Kageyama and Robertson (1993) have described similar features in radial glia of auditory and visual nuclei of neonatal rat and hamster, including the absence of labeling in the dorsal-medial portion of the SC, after labeling of afferent fibers by WGA-HRP.

The lack of cell labeling at and near the midline (MVF) is accompanied by other features that support the notion of a dual system of radial fibers in the opossum and in rodents (Barradas et al., 1988; 1989; Kageyama and Robertson, 1993) and, as judged from other criteria, in several placentals (Van Hartesveldt et al., 1986; Wu et al., 1988; Mori et al., 1990; Snow et al., 1990; Wu et al., 1990). Far from signifying merely the lack of interactions with optic axons, the failure to accumulate axon-borne HRP may have a deep functional significance, that of MVF as a selective barrier to their growth. There is in vivo evidence that the median radial glia of the midbrain tectum may function as a filter that prevents retinal axons from crossing the midline (Wu et al., 1990). An analogous mechanism may operate at the optic chiasm to prevent the restricted set of ipsilateral fibers from crossing the midline (Godement et al., 1990). Furthermore, the growth of processes of lateral or medial midbrain neurons is thwarted if plated onto astrocytes derived from the medial sector of the midbrain but favored if plated onto astrocytes from the lateral sector (Cavalcante et al., 1992; Garcia-Abreu et al., 1994). Moreover, sulfated proteoglycans, present in the roof plate of the spinal cord and midline tectum, inhibit the in vitro outgrowth of processes of dor-

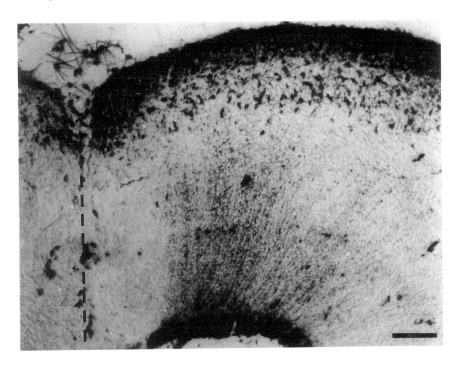

Fig. 2. Filling of radial glia somata (center of lower edge) and inner segments of radial fibers of the SC (superior colliculus) lateral or main radial system (MRS) by axon-borne HRP in a developing opossum (postnatal d 36, postconceptional d 49), 1 wk after an unilateral eye injection. Superficial terminal fields, axons, and/or axonal debris and presumptive macrophages also appear labeled. Interrupted line marks the collicular midline, SC contralateral to the injected eye to the right. Bar, 100 μm.

sal root ganglion cells, a population of neurons that encounters the spinal cord roof plate in vivo (Snow et al., 1990).

With regard to the labeling of the MRS or lateral system, Kageyama and Robertson (1993) compared labeling of radial glia byWGA-HRP and lipophilic carbocyanine dyes and have interpreted the former as owing to *transcellular transport*, i.e., anterograde transport and exocytosis by axons and terminals followed by adsorptive endocytosis (pinocytosis if HRP) by adjacent glial cells and transport toward the cell body, "retrograde glioplasmic transport." Retrograde glioplasmic transport in MRS may not necessarily require *transcellular transport* but may also follow *transcellular diffusion* across contiguous membranes. Irrespective of the exact nature of the antecedent phenomena, it is tempting to speculate that retrograde glioplasmic transport provides a substrate for neuronal messages to radial glia (Barradas et al., 1989; Kageyama and Robertson, 1993). Since a reasonably high number of cells maintains a radial morphology during collicular neurogenesis and for a long while after cellular migration and ingrowth of retinal fibers into the superficial layers, such messages may have different contents in different developmental stages.

An intriguing possibility for axo-(radial)glia interactions in the avian optic tectum has been suggested by Vanselow and coworkers (1989). In the rostral tectum of chick embryos, radial glia endfeet form arrays that are parallel to the orientation of retinal axon fascicles, but such arrangements fail to form after early removal of the optic vesicle. Since the number of endfeet and, presumably, of radial glia cells remains constant, the pattern of endfeet requires interactions between retinal fibers and radial glia to develop.

4.2. The Cytoskeleton of Radial Glial Cells— Vimentin and GFAP

Vimentin immunoreactivity is found everywhere in the SC (Figs. 3–5 and 7) but its highest intensity occurs at and near the midline (MVF), thus, nearly complementary to the distribution of radial processes labeled by axon-borne HRP (Figs. 3 and 4). This denser band has been observed in the hamster (Wu et al., 1988; 1990) as well in the small New World marsupial *Monodelphis domestica* (Elmquist et al., 1994).

The enhanced vimentin immunoreactivity seems to reflect well the ultrastructural features of the MVF, first described in the rat by Raedler and coworkers (1981). These authors also described an interesting feature of the development of MVF, namely, a lateral-to-medial tangential migration within the ventricular zone. We have not investigated this aspect in the opossum SC but it is probably common to all mammals. It is interesting to notice that there is a similar gradient in the generation of neurons destined to the superficial layers both in rat and opossum (Mustari et al., 1979; Cavalcante et al., 1984).

In addition to a presumptive role as a selective barrier to axonal growth across the midline, the MVF may help to form

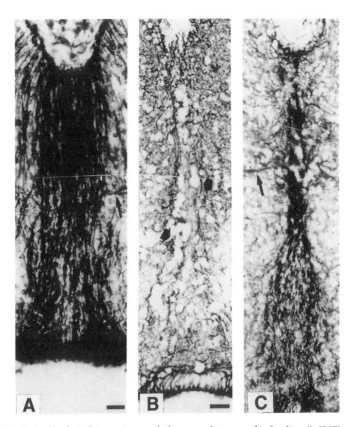

Fig. 3. Cytoskeletal tagging of the median radial glia (MVF, median ventricular formation) in the developing opossum superior colliculus: **A, C.** Vimentin immunoreactivity at postnatal d 15 and 36. Arrows point at a few but constant profiles oriented perpendicularly to the midline. **B.** GFAP-immunoreactivity (large arrow) at postnatal d 31. Bars, 20 µm (A), 30 µm (B = C).

compartments in the midbrain. Lateral protrusions of MVF processes mark the first cytoarchitectonic limits of the rostral midbrain, i.e., the separation of the central gray from the SC proper (*see* arrows in Figs. 3A,C, and 4A).

Both the developing avian optic tectum and mammalian SC display transient patterns of layering before attaining the definite laminar architecture (La Vail and Cowan, 1971; Cavalcante and Rocha-Miranda, 1978b; Raedler et al., 1981; Cooper and Rakic, 1981). Thus, it was hoped that changes in the MRS or lateral system could

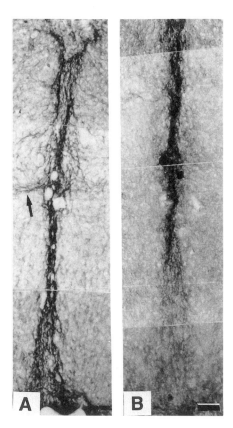

Fig. 4. Vimentin immunoreactivity **(A)** and glycogen deposition **(B)** in MVF in the developing opossum superior colliculus at postnatal d 51 (2 wk before eye opening). Near-adjacent sections, negative image in A of two stained neurons of the mesencephalic trigeminal nucleus in B. Bar, 50 µm.

correlate with transient patterns of layering or predict the appearance of permanent layers. That might be expected for the following reasons:

1. Segments of the permanent radial glia in the amphibian spinal cord exhibit differential immunoreactivity to GFAP as they cross through gray or white matter (Miller and Liuzzi, 1986);
2. In the developing lateral geniculate nucleus of the tree shrew, fine vimentin-reactive processes precede and later become coincident with layers of neuronal cells (Hutchins and Casagrande, 1989).

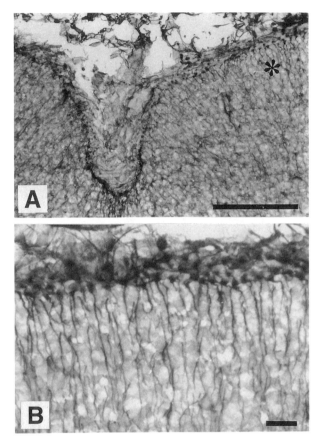

Fig. 5. Vimentin immunoreactivity in MRS in the developing opossum superior colliculus at postnatal d 36. **B** is a magnified view of site marked by asterisk in **A** to show vivid staining of distal segments and endfeet (cf, Fig. 2). Bars, 100 μm (A), 10 μm (B).

No strict demarcation has been observed in the radial axis of the retino-recipient midbrain tectum by the distribution of vimentin (opossum, Barradas et al., 1989; tree shrew, Hutchins and Casagrande, 1989), or of a vimentin-associated antigen (chick, Vanselow et al., 1989), or even the distribution of another precocious marker of radial glia, glutamine synthetase (chick, Linser and Perkins, 1987). In the opossum SC, vimentin-positive radial processes extend without interruption from the subpial endfeet to periacqueductal somata as late as postnatal d 23 (not shown, cf, Fig. 3A in Barradas et al., 1989), i.e., at the time that optic fibers collect into a layer relatively poor in cell bodies (Cavalcante and Rocha-Miranda, 1978a,b). However, in subsequent stages vimentin distribution in MRS undergoes a regression that affects mostly the inner segments (Fig. 4), in spite of filling of these segments with axon-borne HRP in similarly aged animals. That has also been reported for a vimentin-like antigen in the radial glia of the chick optic tectum that becomes restricted to the domain of the (subpial) optic layer, in spite of tagging of the entire cell by a carbocyanine dye in the early posthatching period (Vanselow et al., 1989).

Attempts to label SC radial glia with anti-GFAP antibodies, both monoclonal and polyclonal, are far less successful than with antivimentin antibodies at several stages. That is probably unrelated to the fact that anti-GFAP antibodies were raised against antigens from placental mammals. Not only is GFAP much conserved among vertebrates (Onteniente et al., 1983) but also a recent study in *Monodelphis domestica*, another New World marsupial, has shown that vimentin and GFAP appear as single bands at molecular weights consistent with those reported for other species (Elmquist et al., 1994). However, faint to moderate GFAP-immunoreactivity can be detected in and only in the MVF from postnatal d 12 (by the end of collicular neurogenesis) to postnatal d 36 (by the end of the phase of fast axons loss in the optic nerve) (Fig. 3). In fact, GFAP immunoreactivity persists afterwards in the median region but cannot be ascribed exclusively to radial glia.

4.3. Carbohydrate Deposition
and Other Median-Lateral Differences

Granules of glycogen are often found in astrocytes both of the protoplasmic and the fibrous varieties, being highly uncommon in neurons, oligodendrocytes, or microglia in the mature brain (Peters et al., 1991). This criterion has also been applied to the ultrastructural identification of radial glia (Rakic, 1971) and, at light microscopical level, as a presumptive signal of differentiation of such cells (Schmechel and Rakic, 1979).

Glycogen deposition has proved to be a useful, additional criterion to characterize the midline astroglial system of the SC (Fig. 5). As is the case for GFAP immunoreactivity, this is not an exclusive property of the midline tectal roof, being also found in the midline of the brain stem (tegmental) floor. The midline ensemble is characterized by a multitude of markers that do not appear or are scarce in lateral radial glia and that may range from the calcium-binding protein S100 (Van Hartesveldt et al., 1986) to a soluble cytosolic protein recognized by the R2D5 antibody (Mori et al., 1990). It has also been verified that immunoreactivity to connexin 43, a gap junction protein common to cardiac muscle cells and astrocytes (Dermietzel et al., 1991) also appears earlier in the midline glia than in the remaining SC (Cavalcante et al., 1993).

Contrary to expectations, glycogen accumulation may be a dynamic, regionally selective attribute of astroglia. It has, for instance, been shown that prolonged barbiturate anesthesia causes a prominent accumulation of glycogen in astrocytes of the dentate gyrus and frontal cortex with little, if any, change in other test sites, such as the hypothalamus or midbrain reticular formation (Phelps, 1972). In dissociated cell cultures of the cerebral cortex, a small subset of cortical astrocytes accumulates glycogen (Rosenberg and Dichter, 1987). In both studies, it has been suggested that heterogeneities of glycogen accumulation within an astrocytic population are related to the influences of a glycogenolytic neurotransmitter/neuromodulator, namely, norepinephrine. However, the connection between glycogen accumulation and the definite catecholaminergic innervation of the opossum midbrain (Cosenza and Machado, 1978) remains elusive.

5. The Characteristics of Young Astrocytes

In several regions of the central nervous system, stellate astrocytes are initially vimentin-positive although not necessarily retaining this trait in adult life. In the developing opossum SC, sparse vimentin-positive stellate cell bodies are seen interspersed with radial profiles of the MRS at the level of the superficial layers by postnatal d 26, in correlation with the onset of selective arborization of retino-collicular axons in their prospective target sites (Mendez-Otero et al., 1985; *see* also Simon and O'Leary, 1990). Although few in number, the presence of these vimentin-positive cells exclusively in superficial layers suggests a privileged relationship of the astroglia with optic fibers.

The first GFAP-positive astrocytes in the SC are localized in the superficial layers and are detected by postnatal d 40 (Fig. 6A). One of the conspicuous changes coinciding with the differentiation of astrocytes is the establishment of binocular segregation

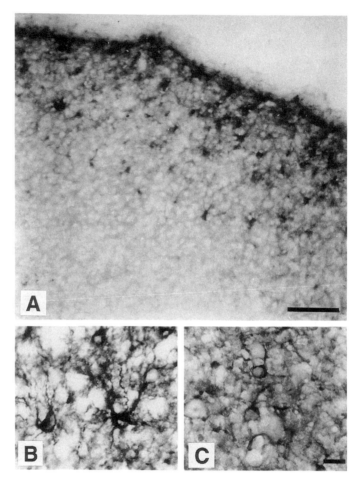

Fig. 6. GFAP immunoreactivity in the developing opossum superior colliculus at postnatal d 40 **(A)**, 46 **(B)**, and 66 **(C)**. A shows the superficial, sharply restricted domain of the first GFAP-positive astrocytes, B and C illustrate the transient enhancement of GFAP-immunoreactivity. Bars, 50 μm (A) and 10 μm (B = C).

of retino-colicular projections with uncrossed terminal fields localized in the zonal layer and the crossed projections in the superficial gray layer (Cavalcante and Rocha-Miranda, 1978a).

It should be noted that postnatal d 40 corresponds to about two-thirds along the cecal period, thus, being nearly equivalent to postnatal d 15 in *Monodelphis domestica* when the first GFAP-positive astrocytes are seen in the midbrain (Elmquist et al., 1994).

The differentiation of astrocytes in the SC, as judged by GFAP-immunoreactivity, follows an outside-in sequence and, thus, departs from the previous developmental gradients, namely, those of neurogenesis (Cavalcante et al., 1984) and cytoarchitectonic differentiation (Cavalcante and Rocha-Miranda, 1978b). Such astrocytes tend to be radially-oriented and coexist with a few radial glia that retain vimentin immunoreactivity in their outer segments and endfeet. These young astrocytes also present the curious characteristic of a transiently high GFAP-immunoreactivity (from postnatal d 40–51, cf Fig. 6B and C), also reported in astrocytes of the mouse optic nerve and tract (Bovolenta et al., 1987) as well as in those of the rat cerebral cortex (LeVine and Goldman, 1988). We have previously suggested that this characteristic in the opossum SC is linked to a progression in the arborizations of optic axons and indeed a study of synaptogenesis therein has shown that along this period there are moderate increases in the numerical density of synapses and other progressive phenomena, such as a clear distinction between retinal and cortical terminals and the first detection of profiles with pleomorphic vesicles (Correa, 1994).

6. Astrocytes and Other Nonneuronal Cells

Glia–glial interactions are often invoked implicitly or overtly in studies of differentiation of astrocytes. One of the best known examples is that of an influence of amoeboid microglia in the proliferation and GFAP expression by astrocytes, mediated by interleukin 1 (Giulian et al., 1988). An analysis of the chronology of the appearance of and changes undergone by microglia in the opossum SC is also in order in view of

1. The recognition that immature microglia might be phagocytic (Lent et al., 1985; Perry et al., 1985; Jordan and Thomas, 1988) and, thus, might express vimentin (Streit et al., 1988);
2. An apparently broad lag between the appearance of vimentin-positive and GFAP-positive cell bodies in the superficial layers. In fact, the presence of amoeboid microglia could be thought of as subjacent to both the elimination of retinal axons (Kirby et al., 1988) and the differentiation of astrocytes.

A study of microglia in the developing opossum SC (Cavalcante and Santos-Silva, 1990) through the binding of the BS1/B4 isolectin of *Griffonia simplicifolia*, a marker of macroph-

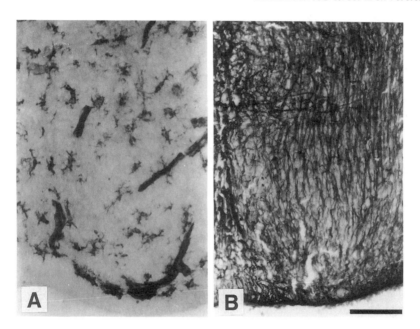

Fig. 7. Microglial marking at the first detection stage **(A)** and vimentin immunoreactivity **(B)** in the developing opossum midbrain at postnatal d 18. Both microglia and blood vessels (thick rods) are labeled by the BS1/B4 isolectin of *G. simplicifolia*. Microglial cells tend to be more ramified in the central gray (lower two-thirds) than in the deep collicular layers (upper third). Bar, 100 μm.

ages/microglia (Streit et al., 1988), showed that soon after the end of collicular neurogenesis, microglial cells appear in the midbrain tectum, first invading the central gray and deep layers of the SC proper (by postnatal d 18, Fig. 7) and later appearing in the superficial layers (by d 26). By the time of appearance of GFAP-positive astrocytes, microglia cells of the superficial layers are fairly ramified (not shown).

The discrepancy between the gradients of appearance of microglia and astrocytes is not readily explained. It cannot be ruled out that microglia have a priming action on neighboring somata of radial glia either lining or adjacent to the lining of the aqueduct so that there would be spatial and temporal discrepancies between their appearance and that of GFAP-positive somata. Alternatively, ramified microglia retain the ability to release factors that promote the proliferation of GFAP-positive cells. The last alternative might explain the sudden appearance of a fairly high number of GFAP-positive cells.

Finally, mention should be made of the spatial and temporal dissociation between the differentiation of astrocytes and oligodendrocytes in the opossum SC (Barradas et al., 1989; Cavalcante et al., 1991). Although there may be a correlation between the transition from vimentin to GFAP and myelination for the entire brain (Dahl, 1981), we have verified that the differentiation of oligodendrocytes follows an outside-in gradient and starts later than the appearance of GFAP-positive astrocytes. Furthermore, the most active phase of myelination in the optic layer is not accompanied by any noticeable change in the number or morphology of astrocytes.

Acknowledgments

This chapter is dedicated to Carlos E. Rocha-Miranda on his 60th birthday. The work reported in and the preparation of this text have been supported by CNPq, FINEP, CEPG/UFRJ, and SR2/UERJ.

References

Achucarro N (1915) De l' évolution de la névroglie, et spécialement de ses relations avec l' appareil vasculaire. Trab Lab Invest Biol 13:169–212.

Allodi S Cavalcante LA Hokoç JN Bernardes RF (1992) Genesis of neurons of the retinal ganglion cell layer in the opossum. Anat Embryol 185:489–499.

Amundson RH Goderie SK Kimelberg HK (1992) Uptake of [3H]serotonin and [3H]glutamate by primary astrocyte cultures. II. Differences in cultures prepared from different brain regions. Glia 6:9–18.

Barradas PC Cavalcante LA Mendez-Otero R Vieira AM (1989) Astroglial differentiation in the opossum superior colliculus. Glia 2:103–109.

Barradas PC Mendez-Otero R Vieira AM Cavalcante LA (1988) Dual populations of radial glial cells in the developing opossum superior colliculus. Soc Neurosci Abstr 14:426.

Behan M (1981) Identification and distribution of retinocollicular terminals in the cat: an electron microscopic autoradiographic analysis. J Comp Neurol 199:1–15.

Bovolenta P Liem RKH Mason CA (1987) Glial filament protein expression in astroglia in the mouse visual pathway. Dev Brain Res 33:113–126.

Campbell CBG Hayhow WR (1972) Primary optic pathways in the duckbill platypus Ornitorynchus anatinus. J Comp Neurol 145:195–208.

Cavalcante LA (1987) Postnatal neurogenesis and the development of neural connections in the visual system of a marsupial. Pontificiae Academiae Scientiarum Scripta Varia 59:1–29.

Cavalcante LA Bernardes RF (1978) Observations on dendritic patterns of neurons in the superficial gray layer in the opossum superior colliculus, in Opossum Neurobiology (Rocha-Miranda CE Lent R, eds.), Academia Brasileira de Ciencias, Rio de Janeiro, pp. 127–136.

Cavalcante LA Rocha-Miranda CE (1978a) Postnatal development of retinogeniculate, retinopretectal and retinotectal projections in the opossum. Brain Res 146:231–248.

Cavalcante LA Rocha-Miranda CE (1978b) Development of the dorsal thalamus and the superior colliculus in the opossum, with special reference to optic projection areas, in Opossum Neurobiology (Rocha-Miranda CE Lent R, eds.), Academia Brasileira de Ciencias, Rio de Janeiro, pp. 193–216.

Cavalcante LA Santos-Silva A (1990) Microglia in the developing opossum superior colliculus. Society Neurosci Abstr 16:176.

Cavalcante LA Barradas PC Martinez AMB (1991) Patterns of myelination in the opossum superior colliculus with additional reference to the optic tract. Anat Embryol 183:273–285.

Cavalcante LA Mendez-Otero R, Allodi S (1985) Interaction between radial glial cells and optic fibers during development. Braz J Med Biol Res 18:639A.

Cavalcante LA Rocha-Miranda CE Linden R (1984) Observations on neurogenesis of the superior colliculus and pretectum in the opossum. Dev Brain Res 13:241–249.

Cavalcante LA Garcia J Carvalho SL Moura Neto V (1992) Sectorial differences in interactions of the mesencephalic glia with local neurons. Int J Dev Neurosci 10 (Suppl. 1): 74.

Cavalcante LA Avanzi D Modesto E Carvalho SL Campos-de-Carvalho AC (1993) Regionally selective distribution of a gap junction protein in median radial glia of the opossum midbrain. Soc Neurosci Abstr 19:60.

Cooper ML Rakic P (1981) Neurogenetic gradients in the superior and inferior colliculi of the rhesus monkey. J Comp Neurol 202:309–334.

Correa EM (1994) Synaptogenesis in retino-recipient layers of the superior colliculus in Didelphis marsupialis (in Portuguese). Ph.D. thesis, Instituto de Biofisica C. Chagas Filho, Universidade Federal do Rio de Janeiro.

Cosenza RM Machado ABM (1978) Catecholamine-containing nerve terminals in the brainstem of the opossum (Didelphis albiventris Lund 1841), in Opossum Neurobiology (Rocha-Miranda CE Lent R, eds.), Academia Brasileira de Ciencias, Rio de Janeiro, pp. 269–288.

Crossland WJ Umchwat CJ (1982) Neurogenesis in the central visual pathways of the golden hamster. Dev Brain Res 5:99–103.

Dahl D (1981) The vimentin-GFA protein transition in rat neuroglia occurs at the time of myelination. J Neurosci Res 6:741–748.

Debski EA Cline HT, Constantine-Paton M (1990) Activity-dependent tuning and the NMDA receptor. J Neurobiol 21:24–32.

De Long GR Sidman RL (1962) Effects of eye removal at birth on histogenesis of the mouse superior colliculus. An autoradiographic analysis with tritiated thymidine. J Comp Neurol 118:205–224.

Dermietzel R Hertzberg EL Kessler JA Spray (1991) Gap junctions between cultured astrocytes: immunocytochemical, molecular, and electrophysiological analysis. J Neurosci 11:1421–1432.

Elmquist JK Swanson JJ Sakaguchi DS Ross LR, Jacobson C (1994) Developmental distribution of GFAP and vimentin in the Brazilian opossum brain. J Comp Neurol 344:283–296.

Garcia-Abreu J Moura Neto V Carvalho SL, Cavalcante LA (1994) Regionally-specific properties of midbrain glia. I. Interactions with midbrain neurons. J Neurosci Res, in press.

Giulian D Young DG Woodward J Brown DC Lachman LB (1988) Interleukin 1 is an astroglial growth factor in the developing brain. J Neurosci 8:709–714.

Godement P Salaun J Imbert M (1984) Prenatal and postnatal development of retino-geniculate and retinocollicular projections in the mouse. J Comp Neurol 230:552–575.

Godement P Salaun J Mason CA (1990) Retinal axon pathfinding in the optic chiasm: divergence of crossed and uncrossed fibers. Neuron 5:173–186.

Guillery RW (1969) The organization of synaptic interconnections in the laminae of the dorsal lateral geniculate nucleus. Z Zellforsch Mikrosk Anat 96:1–38.

Huber GC Crosby EC (1943) A comparison of the mammalian and reptilian tecta. J Comp Neurol 78:133–169.

Huerta MF Harting JK (1984) The mammalian superior colliculus: studies of its morphology and functions, in Comparative Neurology of the Optic Tectum (Vanegas H, ed.), Plenum, New York, pp. 687–773.

Hutchins JB Casagrande VA (1989) Vimentin: Changes in distribution during brain development. Glia 2:55–66.

Innocenti GM Koppel H Clarke S (1983) Transitory macrophages in the white matter of the developing visual cortex. I. Light and electron microscopic characteristics and distribution. Dev Brain Res 11:39–53.

Jordan FL Thomas WE (1988) Brain macrophages: questions of origin and interrelationship. Brain Res Rev 13:165–178.

Kageyama GH Robertson RT (1993) Transcellular retrograde labeling of radial glial cells with WGA-HRP and DiI in neonatal rat and hamster. Glia 9:70–81.

Kirby MA Wilson PD Fischer TM (1988) Development of the optic nerve of the opossum *(Didelphis virginiana)*. Dev Brain Res 44:37–48.

LaVail JH Cowan WM (1971) The development of the chick optic tectum. I. Normal morphology and cytoarchitectonic development. Brain Res 28:391–419.

Lent R Linden R Cavalcante LA (1985) Transient populations of presumptive macrophages in the developing hamster brain, as indicated by endocytosis of blood-borne horseradish peroxidase. Neuroscience 15:1203–1215.

LeVine S Goldman JE (1988) Embryonic divergence of oligodendrocyte and astrocyte lineages in developing rat cerebrum. J Neurosci 8:3992–4006.

Linser PJ Perkins M (1987) Gliogenesis in the embryonic avian optic tectum: neuronal-glial interactions influence astroglial phenotype maturation. Dev Brain Res 31:277–290.

Marshall LG (1979) Evolution of metatherian and eutherian (mammalian) characters: a review based on cladistic methodology. Zool J Linnean Soc 66:369–410.

McCrady E Jr (1938) The Embryology of the Opossum. Wistar Institute, Philadelphia.

Meek J (1983) Functional anatomy of the tectum mesencephali of the goldfish. An explorative analysis of the functional implications of the laminar structural organization of the tectum. Brain Res Rev 6:247–297.

Mendez-Otero R Cavalcante LA Rocha-Miranda CE Bernardes RF Barradas PC (1985) Growth and restriction of the ipsilateral retinocollicular projection in the opossum. Dev Brain Res 18:199–210.

Miller RH Liuzzi FJ (1986) Regional specialization of the radial glial cells of the adult frog spinal cord. J Neurocytol 15:187–196.

Monzón-Mayor M Yanes C Ghandour MS De Barry J Gombos G (1990) Glial fibrillary acidic protein and vimentin immunohistochemistry in the developing and adult midbrain of the lizard Gallotia galloti. J Comp Neurol 295:569–579.

Mori K Ikeda J Hayaishi O (1990) Monoclonal antibody R2D5 reveals midsagittal radial glia system in postnatally developing and adult brainstem. Proc Natl Acad Sci USA 87:5489–5493.

Murray M Edwards MA (1982) A quantitative study of the reinnervation of the goldfish optic tectum following optic nerve crush. J Comp Neurol 209:363–676.

Mustari MJ Lund RD Graubard K (1979) Histogenesis of the superior colliculus in the albino rat: A tritiated thymidine study. Brain Res 164:39–52.

Onteniente B Kimura H Maeda T (1983) Comparative study of the glial fibrillary acidic protein in vertebrates by PAP immunocytochemistry. J Comp Neurol 215:427–436.

Oswaldo-Cruz E Rocha-Miranda CE (1968) The Brain of the Opossum (Didelphis marsupialis), Instituto de Biofisica da Universidade Federal do Rio de Janeiro, Rio de Janeiro.

Perry VH Hume DA Gordon S (1985) Immunohistochemical localization of macrophages and microglia. Neuroscience 15:313–326.

Peters A Palay SL Webster H deF (1991) The Fine Structure of the Nervous System, 3rd ed., Oxford University Press, Oxford.

Phelps CH (1972) Barbiturate-induced glycogen accumulation in brain. An electron microscopic study. Brain Res 39:225–234.

Qian J Bull MS Levitt P (1992) Target-derived astroglia regulate axonal outgrowth in a region-specific manner. Dev Biol 149:278–294.

Raedler E Raedler A Feldhaus (1981) Prenatal differentiation of colliculus superior in the rat. Biblthca Anat 19:174–191.

Raff MC (1989) Glial cell diversification in the rat optic nerve. Science 243: 1450–1455.

Rakic P (1971) Neuron-glia relationship during granule cell migration in developing cerebellar cortex. A Golgi and electron microscopic study in Macacus rhesus. J Comp Neurol 141:283–312.

Ramon y Cajal S (1911) Histologie du systeme nerveux de l'homme et des vertebres, Maloine, Paris, pp. 174–250.

Raymond PA (1986) Movement of retinal terminals in goldfish optic tectum predicted by analysis of neuronal proliferation. J Neurosci 6:2479–2488.

Repérant J Peyrichoux J Rio JP (1981) Fine structure of the superficial layers of the viper optic tectum. A Golgi and electron microscopic study. J Comp Neurol 199:393–417.

Robinson S Dreher B (1990) The visual pathways of eutherian mammals and marsupials develop according to a common timetable. Brain behav evol 36:177–195.

Roots BI (1986) Phylogenetic development of astrocytes, in Astrocytes: Development, Morphology, and Regional Specialization of Astrocytes, vol.1 (Fedoroff S Vernadakis A, eds.), Academic, Orlando, FL, pp. 1–34.

Rosenberg PA Dichter MA (1987) A small subset of cortical astrocytes in culture accumulates glycogen. Int J Dev Neurosci 5:227–235.

Sanderson KJ Aitkin LM (1990) Neurogenesis in the visual and auditory pathways of a marsupial: the brush-tailed possum (Trichosurus vulpecula). Brain Behav Evol 35:325–338.

Schmechel DE Rakic P (1979) A Golgi study of radial glial cells in developing monkey telencephalon: morphogenesis and transformation into astrocytes. Anat Embryol 156:115–152.

Schoenitzer K Hollander H (1984) Retinotectal terminals in the superior colliculus of the rabbit: A light and electron microscopic analysis. J Comp Neurol 223:153–162.

Simon DK O'Leary DDM (1990) Limited topographic specificity in the targeting and of mammalian retinal axons. Dev Biol 137:125–134.

Snow DM Lemmon V Carrino DA Caplan AI Silver J (1990) Sulfated proteoglycans in astroglial barriers inhibit neurite outgrowth in vitro. Exp Neurol 109:111–130.

Stein B (1984) Development of the superior colliculus. Ann Rev Neurosci 7:95–125.

Streit WJ Graeber MB Kreutzberg GW (1988) Functional plasticity of microglia: a review. Glia 1:301–307.

Tigges M Tigges J Luttrell GL Frazier CM (1973) Ultrastructural changes in the superficial layers of the superior colliculus in Galago crassicaudatus. Z Zellforsch mikrosk Anat 140:291–307.

Valentino KL Jones EG (1982) Morphological and immunocytochemical identification of macrophages in the developing corpus callosum. Anat Embryol 163:157–172.

Van Hartesveldt C Moore B Hartman BK (1986) Transient midline raphe glial structure in the developing rat. J Comp Neurol 253:175–184.

Vanselow J Thanos S Godement P Henke-Fahle S Bonhoeffer F (1989) Spatial arrangements of radial glia and ingrowing retinal axons in the chick optic tentum during development. Dev Brain Res 45:15–27.

Wilczynski W Zakon H (1982) Transcellular transfer of HRP in the amphibian visual system. Brain Res 239:29–40.

Wu DY Jhaveri S Moya KL, Schneider GE (1988) Vimentin and GFAP expression in developing hamster superior colliculus. Soc Neurosci Abstr 14:1110.

Wu DY Schneider GE Jhaveri S (1990) Retinotectal axons cross to the wrong side following disruption of the tectal midline cells in the hamster. Soc Neurosci Abstr 16:336.

Neuron/Glia Lineages During Early Nervous System Development in Amphibian and Chicken Embryos

Philippe Cochard, Cathy Soula,
Marie-Claude Giess,
Françoise Trousse, Françoise Foulquier,
and Anne-Marie Duprat

1. Introduction

Neurons and macroglial cells of the vertebrate central nervous system (CNS) arise from neuroepithelial precursor cells that proliferate rapidly in the ventricular and subventricular zones and present similar morphological characteristics. In most CNS regions, neurons develop first, together with a subpopulation of immature glial cells, the radial glia. The other glial cell types, astrocytes and oligodendrocytes, develop later on according to spatio-temporal schemes specific for each central structure. This sequence of differentiation events raises several key questions regarding the lineage relationships of the various CNS cell types and the mechanisms by which these lineages segregate and differentiate. One important aspect of these questions is to define when and how precursor cells become irreversibly committed to a specific differentiation pathway. Neuroepithelial cells initially could be all endowed with equivalent differentiation capabilities. The specification of these multipotential precursor cells toward a defined phenotype could occur progressively, resulting, as development proceeds, in the formation of discrete families of determined

From: *Neuron-Glia Interrelations During Phylogeny: I. Phylogeny and Ontogeny of Glial Cells*
A. Vernadakis and B. Roots, Eds. Humana Press Inc., Totowa, NJ

progenitors. Alternatively, they could be restricted in their developmental fate only when they undergo their final round of division, just before differentiating. In either of these two alternatives, the environment should play an important role in specifying the ultimate cell phenotype. In another possible scheme, neuro-epithelial progenitors could be already segregated into subpopulations with differing potentialities at early stages in nervous system ontogeny. In this case, environmental cues would be less critical than intrinsic developmental programs in regulating phenotypic choices.

Among the various strategies developed to address these questions, cell cultures of defined subsets of progenitors and cell cloning in vivo and in vitro have provided new and important insights into the developmental repertoire of CNS precursor cells. The clonal approach, in particular, provides a direct means of assessing the potential and fate of individual progenitors.

Three techniques are currently available for cloning neural progenitors: retrovirus-mediated transfer of a genetic marker (Price, 1987; Cepko, 1988; Sanes, 1989), intracellular labeling with enzymes such as peroxidase (Holt et al., 1988) or vital fluorescent dyes (Bronner-Fraser and Fraser, 1988) and physical isolation of single CNS blast cells in vitro (Temple and Raff, 1985; Temple, 1989). Each of these techniques has its own advantages and limitations. The retroviral approach, in particular, is a powerful technique, since clonally related cells are permanently labeled, up to adult stages, whereas injected enzymes or fluorescent tracers will dilute as development proceeds as a result of cell division. On the other hand, intracellular labeling allows one to choose precisely the location of the injected cell, whereas the infection of neural progenitors by retroviral particles is an uncontrolled, statistical process. The main drawback in the retroviral approach is that insertion of the retroviral gene occurs in only one of the daughter cells of the dividing progenitor (Hajihosseini et al., 1993). In other words, the observed progeny does not reflect the fate of the infected progenitor, but only that of one of its daughter cells. Therefore, in case of an asymmetrical division, the fate of the other daughter cell will be entirely ignored.

Despite this difficulty in interpreting the data obtained using retroviral markers, the main lesson that can be drawn from experiments using the various clonal strategies is that the segregation of neural cell lineages does not follow a simple and universal scheme throughout the vertebrate CNS. Results obtained so far with respect to the segregation of neuronal and glial lineages are summarized in Table 1. For instance, in the retina of amphibians and rodents, precursor cells are multipotential, generating both neurons and glial cells, until late in development (Turner and Cepko, 1987; Holt et al., 1988; Wetts and Fraser, 1988; Wetts et al., 1989; Turner et al., 1990). In this particular CNS region, therefore, instructive cues from the environment are believed to play a critical role in defining cell phenotypes. Similarly, the early chick embryonic optic tectum and spinal cord contain multipotential precursors, giving rise to mixed clones of neurons and astrocytes (Galileo et al., 1990; Leber et al., 1990; Gray and Sanes, 1991, 1992), which, in the tectum, also include radial glia (Gray and Sanes, 1992). In the E16 cerebral cortex of rodents, mixed clones composed of both neurons and oligodendrocytes have been observed (Williams et al., 1991), but most cloned precursors give rise to homogeneous populations of either neurons, astrocytes, or oligodendrocytes, suggesting that precursors determined to a single phenotype coexist in the ventricular zone (Luskin et al., 1988; Price and Thurlow, 1988). Interestingly, however, elegant in vitro cloning experiments of E14 rat septum reveal that about one-fourth of septal progenitor cells can generate mixed clones composed of both neurons and astrocytes (Temple, 1989). Such a discrepancy between results obtained in in vivo and in vitro experiments discloses a fundamental issue worth mentioning here: Lineage analysis in situ reveals the actual fates of clonally related cells, but not necessarily the full range of potentialities of the ancestor progenitor. Conversely, in appropriate culture conditions, cells cloned in vitro may give rise to progeny exhibiting a much broader range of phenotypes than they would actually develop in vivo. Therefore, results obtained through each strategy give complementary information and cannot be directly compared.

We have chosen to study the segregation of neural lineages in the amphibian embryo. This choice was dictated by several advantages of this model: Neuroepithelial cells are initially of large size and easily accessible at the earliest stages in nervous system ontogeny, i.e., immediately after neural induction. Moreover, the nervous system of the hatched larvae is relatively simple, especially in the posterior rhombencephalon and spinal cord, permitting an easy identification of glial and neuronal cell types. Sections

Table 1
Fate of Neural Precursors in Different Species and in Various CNS Areas[a]

CNS area	Developmental stage[b]	Species	Cloning method	Fate of progenitor[c]	Reference
Rhombencephalon and spinal cord	Neural plate	Pleurodeles	LRD	N/A	(Soula et al., 1993)
Spinal cord	E2-E3	Chick	Retrovirus	N	(Leber et al., 1990)
Optic tectum	E3-E5	Chick	Retrovirus	N/A	(Galileo et al., 1990; Gray and Sanes, 1992)
				N/A/RG	
				N	
				A	
Retina	St. 23	*Xenopus*	LRD	N/M	(Wetts and Fraser, 1988)
	St. 22-27	*Xenopus*	HRP	N/M	(Holt et al., 1988)
	E13-E14	Mouse	Retrovirus	N/M	(Turner et al., 1990)
	Postnatal	Rat	Retrovirus	N/M	(Turner and Cepko 1987)
Cerebral cortex	E12-E14	Mouse	Retrovirus	N	(Luskin et al., 1988)
				A	
Cerebral cortex	E16	Rat	Retrovirus	A	(Price and Thurlow, 1988; Williams et al., 1991)
				O	
				N/O	

[a]Deduced from in vivo experiments using distinct cell cloning strategies.
[b]For chicken and rodents, stages are given in embryonic days (E)
[c]Clones composed of a single cell type are indicated by initials: N neurones; A astroglial cells; O oligodendrocytes; RG radial glia, M Müller cells. In mixed clones, comprising several cell types, initials of the constituting cells are separated by a dash (/). For example, N/A/RG denotes mixed clones composed of neurons, astrocytes and radial glia.

2. and 3. will summarize some of our findings on the potential and fate of early neuroepithelial precursors. Section 4. will be devoted to the analysis of oligodendrocyte potentialities in the chick nervous system and of environmental influences on the expression of the oligodendrocyte phenotype.

2. Expression of Neuron and Astrocyte Phenotypes in Amphibian CNS In Vivo and In Vitro

As a prerequisite to the study of lineage segregation in the amphibian nervous system, we have documented the normal in vivo and in vitro development of astrocytes in amphibians (Soula et al., 1990). *Pleurodeles*, a urodele amphibian, was preferred to the more widely used anuran *Xenopus*, for at least two reasons:

1. Neurulation is much slower in the former species, thus permitting manipulations at critical developmental stages over longer periods of time; and
2. The *Xenopus* neural plate is initially made up of two cell layers, whereas that of *Pleurodeles* is constituted of a single cell layer, and therefore can be better compared to that of higher vertebrates.

2.1. Development of Astrocytes in the Amphibian CNS In Vivo

Since the pioneering work of Bignami et al. (1972), the intermediate filament glial fibrillary acidic protein (GFAP) has been widely recognized in a vast number of species as a highly specific marker for astrocytes and for a limited number of related cell types in the nervous system. We therefore used anti-GFAP antibodies to trace the development of astrocytes in *Pleurodeles* nervous system. On western blots of *Pleurodeles* brain extracts, our antibody stained a band migrating with an apparent molecular mass of 62 kDa, which was identified as a GFAP-like protein (Soula et al., 1990). Thus, *Pleurodeles* astrocytes can be readily recognized by antibodies directed against GFAP (*see also*, Zamora and Mutin, 1988), as it is also the case in other anuran and urodele species, *Xenopus*, the axolotl and *Salamandra* (Godsave et al., 1986; Miller and Liuzzi, 1986; Szaro and Gainer,

1988; Messenger and Warner, 1989; Naujoks-Manteuffel and Roth, 1989).

In *Pleurodeles* (Soula et al., 1990), as in *Xenopus* (Szaro and Gainer, 1988) and in the Axolotl (Messenger and Warner, 1989), expression of GFAP is a precocious event in astrocyte differentiation. The development of GFAP-immunoreactive cells in *Pleurodeles* CNS is schematized in Fig. 1. GFAP is initially detectable as early as at stage 24, i.e., about 2 d after neural plate formation, which in *Pleurodeles* occurs at stage 13. At this stage and until stage 32, astrogliocytes appear as bipolar cells, extending two long radially oriented processes that span the entire thickness of the neural tube wall. Interestingly, at these early stages, immunoreactivity is extremely polarized, initially detectable only at the pial pole of the cell, suggesting an active and oriented transport mechanism of GFAP. Around stage 38, astrocytes undergo a striking morphological change: They become monopolar, with a cell body localized in the ventricular zone of the neural tube, from which a thick radial process emerges, running through the intermediate zone. This unique process then ramifies, within the now well-developed marginal zone, into several thinner processes that end up against the pial surface, forming enlarged and highly immunoreactive endfeet (Fig. 2A). GFAP-immunoreactive cell somata were rarely observed in the gray matter, away from the ventricular zone, thus confirming the conclusions of Naujoks-Manteuffel and Roth (1989) for *Salamandra*, who stated that in urodeles "displaced gliocytes" are a rare occurrence. It is important to mention here that mitotic figures displaying a high level of GFAP immunoreactivity were frequently observed in the ventricular zone (Figs. 1 and 2A), thus suggesting that such radial astrogliocytes are not postmitotic, a property that also seems to apply to the radial glia of higher vertebrates (Misson et al., 1988). In conclusion, most astrocytes in the *Pleurodeles* CNS appear to develop early and retain throughout life the characters of typical radial glia.

2.2. Emergence of Neuron and Astrocyte Phenotypes from Neural Plate Precursor Cells Cultivated In Vitro

Neural plate formation is mediated in vivo by the chorda-mesoderm, through the process of neural induction. The crucial role of the chorda-mesoderm in neural development is not

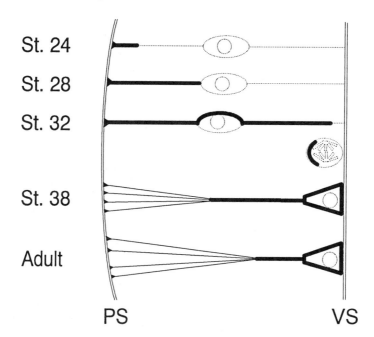

St. 24

St. 28

St. 32

St. 38

Adult

PS   VS

Fig. 1. Schematic drawing recapitulating the localization of GFAP-immunoreactivity in astrocytes, from transverse sections of *Pleurodeles* neural tube at the various developmental stages indicated on the left. GFAP staining is represented as thick and thin solid lines. Dotted lines delineate cell structure. Note that GFAP-immunoreactive mitotic cells, located in the ventricular zone, have been found from stage 30 to 38. *See text* for morphological description. PS: pial surface. VS: ventricular surface. Reprinted with permission from Soula et al., 1990.

restricted to this early step. It has been clearly demonstrated that the notochord is also responsible, after neural tube formation, for the establishment of dorsoventral polarity of the neural tube and for the specification of a number of neuronal types (*see* Placzek et al., 1991, for review). Therefore, it was of interest to determine whether the emergence of astrocytes from neural plate precursor cells would also depend upon epigenetic cues derived from the chorda-mesoderm.

Cells dissociated from stage 13 neural plates were cultivated alone in a simple, completely defined saline medium or cocultivated with chorda-mesoderm cells. Glial cell differentiation was documented at various times by examining the expression of

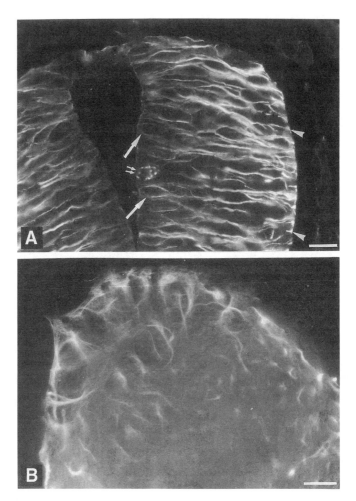

Fig. 2. GFAP immunoreactivity in *Pleurodeles* CNS in vivo **(A)** and in vitro **(B)**. (A) Transverse section at the posterior rhombencephalic level of a stage 34 embryo. Dorsal is at the top. Immunofluorescence delineates cell bodies of astrocytes (arrows) lining the ventricular surface of the neural tube. From these perikarya, numerous intensely stained processes extend radially over the entire width of the neural tube and form enlarged endfeet (arrowheads) against the pial surface of the rhombencephalon. Note the presence of a mitotic glial cell (small arrows) in the ventricular zone, characterized by intense punctated GFAP immunoreactivity. (B) 5-d-old neural plate culture. Numerous stained cell processes are found on top and at the periphery of a large neural aggregate. Bars represent 25 μm.

GFAP. Cells expressing the glial marker developed rapidly in all cultures (Fig. 2B), irrespective of the presence or absence of chorda-mesodermal cells (Soula et al., 1990). This demonstrates that mesodermal cues beyond neural induction are not essentially required for the specification of cells of the astrocytes lineage.

2.3. Neuron and Astrocyte Potentialities Expressed by Cells Dissociated from the Competent Presumptive Ectoderm

It has long been known that neural induction does not strictly depend upon chorda-mesodermal cues, since the presumptive ectoderm can be neuralized without intervention of the natural inducer, by a large number of substances (reviewed in Witkowski, 1985; Saxen, 1989) or in certain experimental situations. In particular, experiments in our laboratory have shown that the dissociation of presumptive ectoderm cells, which occurs readily in Ca^{2+}, Mg^{2+}-free saline medium, is sufficient to provoke the neuralization of a significant fraction of cultivated ectoderm cells as judged from the presence of numerous, morphologically well-differentiated neurons, expressing classical neuronal markers, including neurofilament proteins (Saint-Jeannet et al., 1990). Surprisingly, however, these neurons expressed none of the classical neurotransmitter phenotypic traits, e.g. acetylcholine, GABA catecholamines, and their synthesizing enzymes, and neuropeptides (Saint-Jeannet et al., 1993), which are readily detected in neurons developing from neural plate cultures (Duprat et al., 1985a,b; Pituello et al., 1989a, 1989b, 1990). Therefore, neural induction is not an all-or-none phenomenon, at least in this system, and several inductive steps appear required for the full expression of neuronal phenotypes.

Staining these cultures with anti-GFAP antibodies revealed the differentiation of numerous astrocytes (Saint-Jeannet et al., 1990, 1993), displaying the typical morphology observed in neural plate cultures. Taken together, these results suggest that specification of two of the major neural cell lines, neurons, and astrocytes, can occur totally independently of chorda-mesodermal influences. Experiments in chick embryos, discussed later, indicate that this is probably not true for the development of oligodendrocytes.

3. Clonal Analysis of the Segregation of Neuron and Astrocyte Lineages in the Early Amphibian Neural Plate

The rapid differentiation of neurons and astrocytes in cultures of precursors isolated from the neural plate raised the possibility that at least some of these cells could be, at this early stage, already determined toward restricted neuronal and astrocytes fates. To obtain direct answers to this question, we have labeled individual neural plate precursor cells with a fluorescent tracer and followed the development of their progeny in vivo (Soula et al., 1993). The tracer, lysinated rhodamine dextran (LRD) was injected by iontophoresis as previously described for neural crest and retinal lineage studies (Bronner-Fraser and Fraser, 1988, 1989; Wetts and Fraser, 1988; Wetts et al., 1989).

About 120 cells were labeled at the neural plate stage in the presumptive area of the rhombencephalon and of the spinal cord, and 15 were injected during neural tube closure in the trunk neural tube. Analysis of these clones was carried out at different developmental stages, up to stage 38, i.e., about 12 days after injection. Owing to a rather slow rate of division of neural tube cells in *Pleurodeles* embryos (Chibon, 1973), clone sizes never exceeded 24 cells, a number representing a maximum of five mitoses. Most clones at stages 34–38 contained between six and 16 cells (maximum three to four mitoses). Therefore, the concentration of the tracer within clonally related cells was sufficiently high to permit their easy detection. Neuronal and astrocyte phenotypes of labeled cells were deduced from morphological, positional, and immunocytochemical criteria.

The main finding obtained during this study was that most (80%) neural plate and early neural tube precursors are bipotential, i.e., they give rise in vivo to mixed clones, constituted of both neurons and astrocytes (Fig. 3). In some cases, clones contained only these two cell types, but generally some of the clonally related cells could not be positively identified. Therefore, it is possible that such clones contained other cell types, including undifferentiated precursors and oligodendrocytes. However, not all clones were bipotential. A small but significant number (20%) of clones were strictly constituted of neurons (Fig. 4). Finally, clones constituted exclusively of astrocytes were never observed.

Since the position of each labeled cell in the neural plate was precisely recorded at the time of injection, a fate map was reconstructed (Fig. 5). It was very striking to observe that although precursors for mixed clones were distributed homogeneously in the neural plate, those giving rise to neuronal clones were not. They were never found in the medial and lateral regions of the neural plate, but were exclusively located along the intermediate axes.

This lineage study thus reveals that in urodele amphibians, most early neural plate precursors generate both neurons and astrocytes. The remaining ones, which are not randomly distributed, have a restricted neuronal fate. The significance of such a restriction in fate and in space remains unknown at the present time. In any case, if our observations can be generalized to other species (which remains to be established), the coexistence, within the CNS ventricular zone, of precursor cells with varying fates, already described for later developmental stages (*see* Table 1), may be traced back to the earliest stage in nervous system ontogeny. But the important question raised by this study is whether such an early heterogeneity in cell fate is the result of a similar heterogeneity in the determination of neural progenitors. Alternatively, all precursors could be bipotential at the origin, some becoming restricted to a neuronal fate later on by environmental cues. To answer this question, we are currently developing an in vitro clonal assay system that will allow us to follow the potentialities of individual neural plate precursor cells grown in the most permissive environmental conditions. This assay will then be used to study the fate of progenitors cultivated in various environmental conditions and in particular in the presence of defined growth and differentiation factors.

4. Environmental Influences on the Expression of the Oligodendrocyte Phenotype

The lineage experiments described and discussed in the above sections suggest that early in development, at least some CNS precursor cells may have a restricted developmental repertoire. The important question, therefore, is to understand the mechanisms responsible for the restriction of their differentiation potentialities and controlling their ultimate fate *in situ*. Our own experiments in the amphibian neural plate indicate that cells with

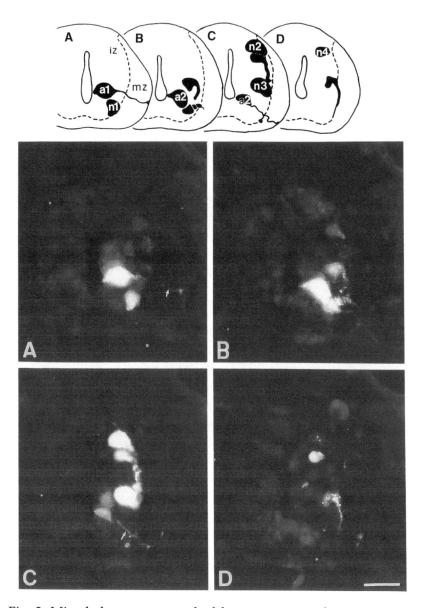

Fig. 3. Mixed clone, composed of four neurons and two astrocytes, observed on four consecutive transverse sections of the spinal cord at stage 38 (dorsal is at the top). The clonally related cells are visualized by the bright fluorescence of LRD. Astrocytes (a1 and a2 on the schematic reconstitution above) are easily recognized in **(A)** and **(B)** by their ovoid cell bodies located against the ventricular surface. Note in (A) and **(C)** the radial processes (arrows), typical of astrocytes, extended,

a restricted neuronal fate are not randomly distributed. Other types of progenitors with a limited repertoire may also be precisely located. In this section, we will review work performed in the oligodendrocyte lineage that establishes that oligodendrocyte precursors are not uniformly distributed in the CNS but restricted into defined CNS areas. Such a precise distribution may reveal the existence of specific environmental conditions prevailing in these areas and responsible for the determination of these precursors.

4.1. Oligodendrocyte Potentialities in the Optic Nerve Primordium and in Other CNS Regions

One of the best characterized CNS progenitor is the O-2A cell, a bipotential neural precursor generating in culture oligodendrocytes and a subclass of astrocytes (*see* Raff, 1989, for a review). O-2A cells were initially identified in the developing rat optic nerve (Raff et al., 1983), but there is now evidence that similar cells are also present in several CNS regions (Levi et al., 1986; Behar et al., 1988). It has been demonstrated clearly that O-2A cells generate oligodendrocytes in the CNS *in situ*, but there are still some doubts regarding the actual generation of astrocytes by O-2A cells in vivo (Skoff, 1990; Skoff and Knapp, 1991). In vitro studies have brought important insights into the mechanisms controlling mitosis, survival and phenotypic choices in the O-2A lineage (*see* Raff, 1989, for a review, and also Barres et al., 1992, 1993; Mayer et al., 1994). In addition, they have revealed that the O-2A cell exhibits extensive migratory capabilities (Small et al., 1987), a property that is shared *in situ* by immature (Gumpel et al., 1983) and mature (Wolf et al., 1986) oligodendrocytes. Investigations on the origin of O-2A cells in the optic nerve have suggested that they may not originate from the optic nerve itself, but rather migrate into the nerve from extrinsic sources (Small et al., 1987). These results have led to the assumption that the developmental

---

respectively, by a1 and a2. Two neurons (n2, n3) can be identified in (C) by a rounded soma apposed against the limit separating the intermediate zone (iz) and marginal zone (mz). The cell process in (D) probably represents an axon extended by neuron n3. Neurons n1 and n4 are only partially included in this series of sections. Bar represents 25 μm.

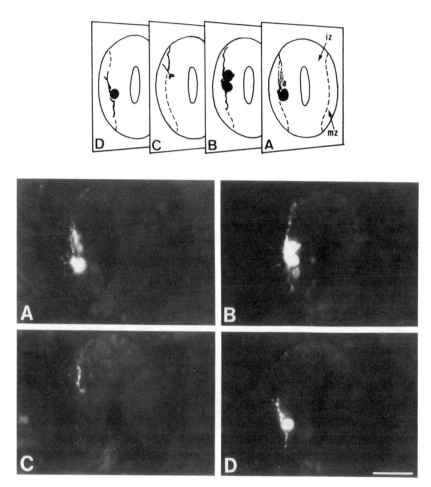

Fig. 4. Clone composed of four neurons analyzed at larval stage 34 on four consecutive transverse sections of the spinal cord. Each section has been schematized above to visualize the rostrocaudal organization of the clone. Each of the four neurons displays a rounded cell body located against the internal border of the marginal zone (mz). Cell bodies, as well as tangentially oriented cell processes, are filled with LRD. Bar represents 25 μm. Reprinted with permission from Soula et al., 1993.

repertoire of cells intrinsic to the embryonic optic nerve may be limited to the production of type-1 astrocytes (Small et al., 1987).

We have reinvestigated this question in the chick embryo, using explant cultures of optic stalks carefully microdissected at various developmental stages (Giess et al., 1990, 1992). Neurons,

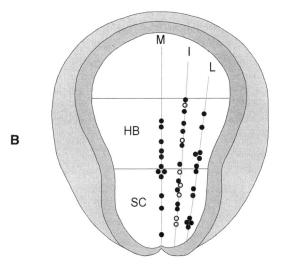

Site of injection	Mixed clones (●)	Neuronal clones (O)
Medial	12/12	0/12
Intermediate	10/17	7/17
Lateral	13/13	0/13

Fig. 5. Fate map of mixed and neuronal progenitors in the neural plate (stage 13), deduced from the coordinates of 42 LRD-labeled cells, recorded at the time of injection, and from phenotypic analysis performed at stages 34 and 38. **(A)** Roughly similar numbers of cells were analyzed along the medial (n = 12), intermediate (n = 17), and lateral (n = 13) axes. **(B)** The fate map clearly shows the exclusive origin of neuronal precursors along the intermediate axis (I) of the neural plate, whereas mixed progenitors were found in all three areas considered. For the purpose of clarity, all progenitors have been represented on only one side of the neural plate, although they were injected on either side. Note that neuronal progenitors were found in presumptive regions of hindbrain (HB) and spinal cord (SC). Reprinted with permission from Soula et al., 1993.

astrocytes, and oligodendrocytes were identified using antibodies directed against neurofilaments, GFAP and galactocerebrosides (GalC), respectively. In addition, Ol1, a monoclonal antibody directed against sulfatides, was used to label immature oligoden-

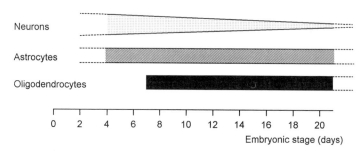

Fig. 6. Summary of neural potentialities in the cultivated embryonic chick optic nerve. Optic stalk cells explanted in vitro at all stages can differentiate into neurons and astrocytes. Neuronal potentialities decrease, however, as development proceeds. In contrast, optic stalk cells from E4 to E6 embryos are unable to generate oligodendrocytes. The oligodendrocytes potential appears abruptly at E7 in the optic nerve.

drocytes (Ghandour and Nussbaum, 1990). Unfortunately, the A2B5 antibody, which recognizes tetrasialogangliosides on the surface of neurons and O2-A cells (Eisenbarth et al., 1979), was not a reliable marker for oligodendrocyte precursors in chick embryos.

Results of these experiments are summarized in Fig. 6. As expected, optic nerve cells gave rise to astrocytes throughout development. In addition, they also appeared capable of generating neurons (Fig. 7A), although this potential diminished progressively during embryonic development (Giess et al., 1990). A similar potential for neuron generation was also demonstrated in the fetal (Giess et al., 1990) and newborn (Omlin and Waldmeyer, 1989) rat optic nerve. Consequently, although the optic nerve is one of the few CNS areas that in the adult are entirely devoid of neuronal cell bodies, neuronal potentialities are initially present in this neuroepithelial territory. Surprisingly, however, optic stalks explanted between 4 and 6 d of incubation (E4–E6) were totally unable to produce oligodendrocytes, even after more than 3 wk in vitro. In contrast, numerous immature (Ol1+/GalC-) and mature (GalC+) oligodendrocytes were invariably observed in cultures from E7 and older optic nerves (Giess et al., 1992).

These experiments indicate that in chick, as in rats (Small et al., 1987), up to a precise developmental stage (E7 in chicks, E17 in rats), optic stalk cells are unable to generate oligodendrocytes. At least two hypotheses can be forwarded to explain these results.

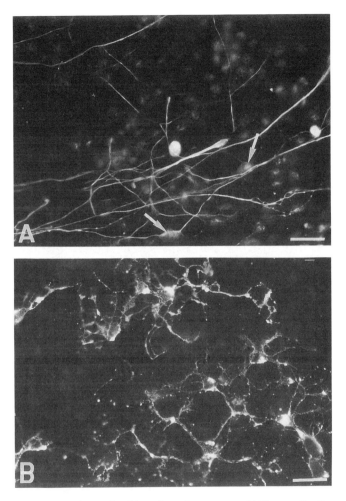

Fig. 7. Neuronal and oligodendrocytes differentiation in the embryonic chick optic nerve. **(A)** E5 optic stalk explant cultivated for 10 d and stained with an antibody directed against neurofilaments. Faintly stained neuronal cell bodies (arrows) extend a network of immunoreactive neurites. **(B)** E7 optic stalk explant cultivated for 4 d, stained with an antibody directed against galactocerebrosides. The antibody decorates numerous well differentiated oligodendrocytes. Bars represent 25 μm.

A first possibility is that embryonic optic stalk precursors are intrinsically devoid of oligodendrocyte potentialities. As proposed by Small et al. (1987), oligodendrocyte precursors populating the optic nerve after E17 in rats, or after E7 in chicks, thus may be of

Table 2
Oligodendrocyte Differentiation in Explant Cultures
of E4 Dorsal and Ventral Spinal Cord[a]

Time in culture, d	Dorsal spinal cord		Ventral spinal cord	
	Ol-1	GalC	Ol-1	GalC
7	0 (4)	0 (4)	>500 (12)	96 ± 35 (12)
15	23 ± 5 (19)	0 (12)	>1000 (6)	>500 (5)

[a]Numbers represent the mean ± SEM of immunoreactive cells per coverslip. Numbers of coverslip examined are indicated in parentheses. Above 200 immunoreactive cells, cell counts were estimated.

extrinsic origin. Alternatively, optic stalk precursor cells may receive, around E6-E7 in chick embryos, a developmental signal of critical importance for their subsequent orientation toward the oligodendrocyte phenotype.

4.2. Oligodendrocyte Potentialities in the Spinal Cord and in Other CNS Regions

An apparent restriction in oligodendrocyte potentialities recently has also been evidenced in the spinal cord. Experiments in which the ventral and dorsal halves of the embryonic (E14) rat spinal cord are cultivated separately indicate that oligodendrocyte potentialities are initially almost entirely restricted to the ventral part of the spinal cord (Warf et al., 1991; Noll and Miller, 1993). Furthermore, at this precise developmental stage, cells expressing the α receptor for platelet-derived growth factor (PDGF-αR), which, in the CNS is expressed by cells of the oligodendrocyte lineage (Pringle et al., 1992) are localized only in the ventral part of the spinal cord ventricular zone (Pringle and Richardson, 1993), thus suggesting a ventral localization of oligodendrocyte precursors.

These results prompted us to analyze oligodendrocyte potentialities in the chick spinal cord as well as in other CNS regions. Dorsal and ventral microexplants were dissected from E4 and E7 embryos and cultivated separately. Cultures were analyzed at various times using the anti-sulfatide (Ol1) and anti-GalC antibodies. Results, summarized in Table 2, indicate that E4 dorsal cord tissue produced few or no oligodendrocytes (Fig. 8A),

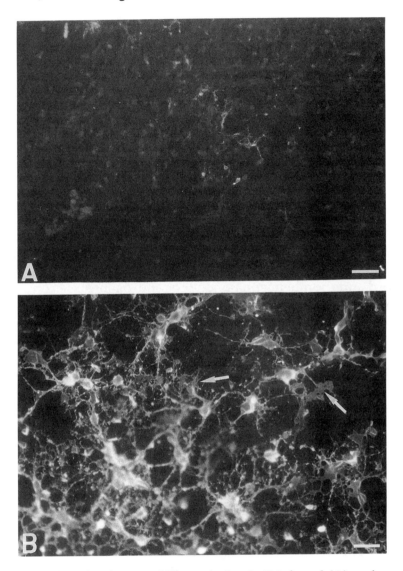

Fig. 8. Oligodendrocyte differentiation in E4 dorsal **(A)** and ventral **(B)** spinal cord explants. Ol1 immunoreactivity in explants cultivated for 15 d. (A) Very few Ol1+ oligodendrocytes are found at the periphery of a dorsal explant. Such cells were very rarely encountered in dorsal cultures and they never expressed galactocerebrosides. (B) In marked contrast, numerous Ol1+ oligodendrocytes were found throughout ventral cultures. At this time in culture, they display a well differentiated morphology, characterized by the extension of membraneous sheets (arrows). Bars represent 50 μm (A) and 25 μm **(C)**.

whereas well-differentiated oligodendrocytes were already identifiable in 7-d cultures of ventral explants (Fig. 8B). This difference in oligodendrogenesis between the dorsal and ventral spinal cord regions is transient and restricted to an early stage in spinal cord ontogeny, since E7 dorsal explants produced numerous oligodendrocytes. Similar results were obtained in cultures of E4 and E7 dorsal and ventral rhombencephalon, indicating that a ventral origin for oligodendrocyte precursors is not limited to the spinal cord. In contrast, the entire E4 mesencephalon, dorsal and ventral, was unable to generate oligodendrocytes in similar culture conditions.

4.3. Notochord Effect on Expression of the Oligodendrocyte Phenotype

The marked ventral predominance in oligodendrocyte potentialities summarized earlier could indicate that in the spinal cord and rhombencephalon, oligodendrogenesis is controlled by ventrally-derived cues, and, in particular, by signals produced by the notochord, an axial mesodermal structure that extends along the ventral aspect of the rhombencephalic and truncal neural tube. The establishment of the dorsoventral polarity of the neural tube is controlled, at least in part, by diffusible signals produced by the notochord and by a specialized subset of cells, located along the ventral midline of the neural tube, termed the floor plate (*see* Placzek et al., 1991, for review). To test this possibility, E4 dorsal explants were cocultivated with segments of notochord dissected from E2.5 and E4 chick embryos. In 80% of cases ($n = 25$), notochord tissue stimulated expression of the oligodendrocyte phenotype (Fig. 9), as evidenced by the differentiation of numerous (from 50 to several hundreds) Ol1+/GalC+ oligodendrocytes, whereas dorsal explants cultivated alone in control experiments contained few, if any, Ol1+ cells and no GalC-immunoreactive cell. However, the effect of notochord tissue was of short range, since oligodendrocytes were produced only by dorsal explants located in contact with the notochord, or in its immediate vicinity (Fig. 9). It was then of interest to investigate the effect of notochord tissue on oligodendrogenesis in the E4 optic stalk. Using the same methodological approach, optic stalk explants were associated in culture with notochord segments. In all cases ($n = 12$), these cocultures were entirely negative for Ol1 and GalC immunoreactivity.

These results have two important implications:

1. E4 dorsal spinal cord tissue, but not E4 optic stalk, contains cells with the ability, upon appropriate stimulation, to produce oligodendrocytes; whether oligodendrocyte precursors that populate later on this CNS region are induced locally, between E4 and E7, by an appropriate developmental signal, or arise from determined progenitors migrating from the ventral cord, as suggested by the group of Miller (Warf et al., 1991; Noll and Miller, 1993) remains to be determined; and
2. Notochord can promote expression of the oligodendrocyte phenotype in the dorsal spinal cord. This raises the interesting possibility that oligodendrocyte determination in the ventral spinal cord could be elicited, directly or indirectly, by the notochord, as already evidenced for a number of neuronal types (Placzek et al., 1991).

In any case, several points still await further study, before definitive conclusions can be drawn about the role of the notochord in this process in vivo. In particular, it will be important to define the specificity of the notochord effect, and the influence of early notochordectomy on oligodendrocyte differentiation in the ventral spinal cord region. It will also be necessary to determine the mechanisms involved in notochord stimulation of oligodendrogenesis.

5. Conclusion

Experiments described in the first part of this chapter clearly indicate that the segregation of neuronal and astrocytes lineages is not an early event in CNS ontogeny. In the amphibian neural plate, even if some cells have a restricted neuronal fate, most precursors generate neurons and astrocytes and therefore are at least bipotential. Iontophoresis injections of a fluorescent tracer in ventricular cells at progressively older developmental stages will help elucidate the lineage relationships of neurons and astrocytes. A similar analysis will also be carried out for the oligodendrocytes lineage, using specific oligodendrocyte markers in amphibians (Steen et al., 1989). In particular, it will be of interest to determine the possible existence in amphibians of a common progenitor for both neurons and oligodendrocytes, as

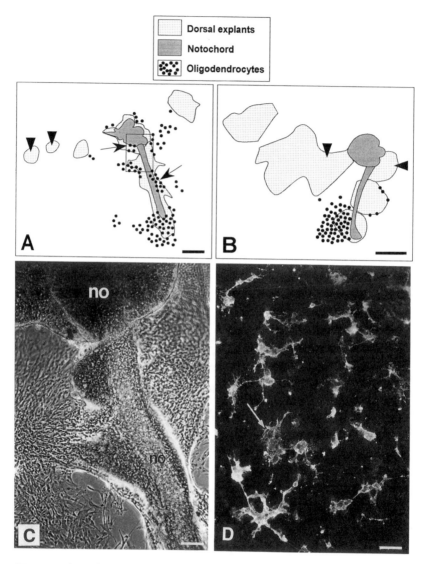

Fig. 9. Oligodendrocyte differentiation after 16 d in E4 dorsal spinal cord/notochord cocultures. **(A, B)** Camera lucida drawings of the areas containing the notochord segments in two such cultures. In each case, one segment has rounded up into a bulgy structure, whereas another remained elongated on the culture substratum. Regions containing well differentiated (Ol1+/GalC+) oligodendroyctes are indicated by black dots. Note that in both cases, most oligodendrocytes were located in close vicinity to notochord tissue. Some were even found in direct contact with the notochord (arrows in A). In contrast, immunoreactive cells were

suggested by retroviral cloning experiments in the rat cerebral cortex in vivo and in vitro (Williams et al., 1991). In addition, in vitro clonal experiments will allow us to determine the full range of potentialities of neural plate precursor cells.

This in vitro approach may also give important insights into the mechanisms regulating lineage decisions and phenotypic choices. Bulk cultures of isolated neural plate or presumptive ectoderm cells have demonstrated that signals produced by the chorda-mesoderm are not decisive in specifying the astrocyte phenotype. Preliminary results obtained in clonal cultures of neural plate cells (Trousse, et al., unpublished data) suggest that the proportion of neuronal clones over mixed neuronal/astrocytesl clones is much greater in vitro than that observed in vivo, indicating that cell differentiation does not follow a rigid program, imposed early in development. The decision for a bipotential progenitor to become a neuron or an astrocyte may therefore be controlled among neuroepithelial cells themselves, through cellular interactions that remain to be defined. One these control mechanisms may involve cell–cell communication through a variety of signalling molecules, including a variety of growth and differentiation factors.

In contrast, expression of the oligodendrocyte phenotype may be induced experimentally in the dorsal spinal cord by the notochord, suggesting that notochord-derived factors may be of importance in defining the location and timing of oligodendrocyte determination. One of them may be the morphogen retinoic acid, synthesized in the notochord and in the ventrally located floor plate (Wagner et al., 1990). Paradoxically, retinoic acid does

---

not observed around dorsal explants located away from notochord tissue (arrowheads in A). However, some dorsal explants developing in contact with the notochord (arrowheads in B) did not produce oligodendrocytes. **(C)** Phase contrast micrograph of the area outlined by the rectangle in A. Notochord segments (no) can be easily recognized by virtue of their three-dimensional structure, whereas the surrounding dorsal spinal cord tissue flattened on the substratum. **(D)** Ol1 immunoreactivity expressed by oligodendrocytes developing in the culture schematized in B. Some cells are well differentiated, as judged from the extension of membraneous sheets (arrow). Bars represent 500 μm (A B), 100 μm (C), and 25 μm (D).

not induce oligodendrocyte differentiation, but conversely inhibits the maturation of A2B5-positive oligodendrocyte precursors (Noll and Miller, 1994). Therefore, other factors may be involved in specifying oligodendrocyte determination. Cells of the O-2A lineage are sensitive to a growing number of soluble factors, which affect selectively, or pleiotropically, their proliferation, differentiation, and survival. Future experiments, using cultures of dorsal spinal cord cells, will help defining environmental factors responsible in the specification of the oligodendrocyte lineage.

References

Barres BA Hart IK Coles HSR Burne JF Voyvodic JT Richardson WD Raff MC (1992) Cell death and control of cell survival in the oligodendrocyte lineage. Cell 70:31–46.

Barres BA Schmid R Sendtner M Raff M (1993) Multiple extracellular signals are required for long-term oligodendrocyte survival. Development 118:283–295.

Behar T McMorris FA Novotny EA Barker JL Dubois-Dalcq M (1988) Growth and differentiation properties of O-2A progenitors purified from rat cerebral hemispheres. J Neurosci Res 21:168–180.

Bignami A Eng LF Dahl D Uyeda CT (1972) Localization of the glial fibrillary acidic protein in astrocytes by immunofluorescence. Brain Res 43:429–435.

Bronner-Fraser M Fraser SE (1988) Cell lineage analysis reveals multipotency of some avian neural crest cells. Nature 335:161–164.

Bronner-Fraser M Fraser SE (1989) Developmental potential of avian trunk neural crest cells in situ. Neuron 3:755–766.

Cepko C (1988) Retrovirus vectors and their applications in neurobiology. Neuron 1:345–353.

Chibon P (1973) Cell proliferation in late embryos and young larvae of the newt *Pleurodeles waltlii* Michah, in The Cell Cycle in Development and Differentiation (Balls M Billett FS, eds.), Cambridge University Press, London/New York, pp. 257–277.

Duprat AM Kan P Foulquier F Weber M (1985a) In vitro differentiation of neuronal precursor cells from amphibian late gastrulae: morphological, immunocytochemical studies, biosynthesis, accumulation and uptake of neurotransmitters. J Embryol Exp Morphol 86:71–87.

Duprat AM Kan P Gualandris L Foulquier F Marty J Weber M (1985b) Neural induction: embryonic determination elicits full expression of specific neuronal traits. J Embryol Exp Morphol 89:167–183.

Eisenbarth GS Walsh FS Nirenberg M (1979) Monoclonal antibody to a plasma membrane antigen of neurons. Proc Natl Acad Sci USA 76:4913–4917.

Galileo DS Gray GE Owens GC Majors J Sanes JR (1990) Neurons and glia arise from a common progenitor in chicken optic tectum: demonstration with two retroviruses and cell type-specific antibodies. Proc Natl Acad Sci USA 87:458–462.

Ghandour MS Nussbaum JL (1990) Oligodendrocyte cell surface recognized by a novel monoclonal antibody specific to sulfatide. Neuroreport 1:13–16.

Giess MC Cochard P Duprat AM (1990) Neuronal potentialities of cells in the optic nerve of the chicken embryo are revealed in culture. Proc Natl Acad Sci USA 87:1643–1647.

Giess MC Soula C Duprat AM Cochard P (1992) Cells from the early chick optic nerve generate neurons but not oligodendrocytes in vitro. Dev Brain Res 70:163–171.

Godsave SF Anderton BH Wylie CC (1986) The appearance and distribution of intermediate filament proteins during differentiation of the central nervous system, skin and notochord of *Xenopus laevis*. J Embryol Exp Morphol 97:201–223.

Gray GE Sanes JR (1991) Migratory paths and phenotypic choices of clonally related cells in the avian optic tectum. Neuron 6:211–225.

Gray GE Sanes JR (1992) Lineage of radial glia in the chicken optic tectum. Development 114:271–283.

Gumpel M Baumann N Raoul M Jacque C (1983) Survival and differentiation of oligodendrocytes from neural tissue transplanted into new-born mouse brain. Neurosci Lett 37:307–311.

Hajihosseini M Iavachev L Price J (1993) Evidence that retroviruses integrate into post-replication host DNA. Embo J 12:4969–4974.

Holt CE Bertsch TW Ellis HM Harris WA (1988) Cellular determination in the *Xenopus* retina is independent of lineage and birth date. Neuron 1:15–26.

Leber SM Breedlove SM Sanes JR (1990) Lineage, arrangement, and death of clonally related motoneurons in chick spinal cord. Neuroscience 10:2451–2462.

Levi G Gallo V Ciotti MT (1986) Bipotential precursors of putative fibrous astrocytes and oligodendrocytes in rat cerebellar cultures express distinct surface features and "neuron-like"-aminobutyric acid transport. Proc Natl Acad Sci USA 83:1504–1508.

Luskin MB Pearlman AL Sanes JR (1988) Cell lineage in the cerebral cortex of the mouse studied in vivo and in vitro with a recombinant retrovirus. Neuron 1:635–647.

Mayer M Bhakoo K Noble M (1994) Ciliary neutrophic factor and leukemia inhibitory factor promote the generation, maturation and survival of oligodendrocytes in vitro. Development 120:143–153.

Messenger NJ Warner AE (1989) The appearance of neural and glial cell markers during early development of the nervous system in the amphibian embryo. Development 107:43–54.

Miller RH Liuzzi FJ (1986) Regional specialization of the radial glial cells of the adult frog spinal cord. J Neurocytol 15:187–196.

Misson JP Edwards MA Yamamoto M Caviness VS (1988) Mitotic cycling of radial cells of the foetal murine cerebral wall: a combined autoradiographic and immunocytochemical study. Dev Brain Res 38:183–190.

Naujoks-Manteuffel C Roth G (1989) Astroglial cells in a salamander brain (*Salamandra salamandra*) as compared to mammals: a glial fibrillary acidic protein immunohistochemistry study. Brain Res 487:397–401.

Noll E Miller RH (1993) Oligodendrocyte precursors originate at the ventral ventricular zone dorsal to the midline region in the embryonic rat spinal cord. Development 118:563–573.

Noll E Miller RH (1994) Regulation of oligodendrocyte differentiation: a role for retinoic acid in the spinal cord. Development 120:649–660.

Omlin FX Waldmeyer J (1989) Differentiation of neuron-like cells in cultured rat optic nerves: a neuron or common neuron-glia progenitor? Dev Biol 133:247–253.

Pituello F Deruntz P Pradayrol L Duprat AM (1989a) Peptidergic properties expressed in vitro by embryonic neuroblasts after neural induction. Development 105:529–540.

Pituello F Kan P Geffard M Duprat AM (1989b) Initial GABAergic expression in embryonic neuroblasts after neural induction of newborn rat. Int J Dev Biol 33:445–453.

Pituello F Boudannaoui S Foulquier F Duprat AM (1990) Are neuronal precursor cells committed to coexpress different neuroactive substances in early amphibian neurulae? Cell Diff Dev 32:71–82.

Placzek M Yamada T Tessier-Lavigne M Jessell TM Dodd J (1991) Control of dorsoventral pattern in vertebrate neural development: induction and polarizing properties of the floor plate. Development 113 (Suppl. 2):105–122.

Price J (1987) Retroviruses and the study of cell lineage. Development 101:409–419.

Price J Thurlow L (1988) Cell lineage in the rat cerebral cortex: a study using retroviral-mediated gene transfer. Development 104:473–482.

Pringle NP Richardson WD (1993) A singularity of PDGF alpha-receptor expression in the dorsoventral axis of the neural tube may define the origin of the oligodendrocyte lineage. Development 117:525–533.

Pringle NP Mudhar HS Collarini EJ Richardson WD (1992) PDGF receptors in the rat CNS - During late neurogenesis, PDGF alpha-receptor expression appears to be restricted to glial cells of the oligodendrocyte lineage. Development 115:535–551.

Raff MC (1989) Glial cell diversification in the optic nerve. Science 243:1450–1455.

Raff MC Miller RH Noble M (1983) A glial progenitor cell that develops *in vitro* into an astrocyte or an oligodendrocyte depending on culture medium. Nature 303:390–395.

Saint-Jeannet JP Huang S Duprat AM (1990) Modulation of neural commitment by changes in target cell contacts in *Pleurodeles waltl*. Dev Biol 141:93–103.

Saint-Jeannet JP Pituello F Huang S Foulquier F Duprat AM (1993) Experimentally provoked neural induction results in an incomplete expression of neuronal traits. Exp Cell Res 207:383–387.

Sanes JR (1989) Analysing cell lineage with a recombinant retrovirus. Trends Neurosci 12:21–28.

Saxen L (1989) Neural induction. Int J Dev Biol 33:21–48.

Skoff RP (1990) Gliogenesis in rat optic nerve - Astrocytes are generated in a single wave before oligodendrocytes. Dev Biol 139:149–168.

Skoff RP Knapp PE (1991) Division of astroblasts and oligodendroblasts in postnatal rodent brain—Evidence for separate astrocyte and oligodendrocyte lineages. Glia 4:165–174.

Small RK Riddle P Noble M (1987) Evidence for migration of oligodendrocyte-type-2 astrocyte progenitor cells into the developing rat optic nerve. Nature 328:155–157.

Soula C Sagot Y Cochard P Duprat AM (1990) Astroglial differentiation from neuroepithelial precursor cells of amphibian embryos: an *in vivo* and *in vitro* analysis. Int J Dev Biol 34:351–364.

Soula C Foulquier F Duprat AM Cochard P (1993) Lineage analysis of early neural plate cells: cells with purely neuronal fate coexist with bipotential neuroglial progenitors. Dev Biol 159:196–207.

Steen P Kalghatgi L Constantine-Paton M (1989) Monoclonal antibody markers for amphibian oligodendrocytes and neurons. J Comp Neurol 289:467.

Szaro BG Gainer H (1988) Immunocytochemical identification of non-neuronal intermediate filament proteins in the developing *Xenopus laevis* nervous system. Dev Brain Res 43:207–224.

Temple S (1989) Division and differentiation of isolated CNS blast cells in microculture. Nature 340:471–473.

Temple S Raff MC (1985) Differentiation of a bipotential glial progenitor cell in single cell microculture. Nature 313:223–225.

Turner DL Cepko CL (1987) A common progenitor for neurons and glia persists in rat retina late in development. Nature 328:131–136.

Turner DL Snyder EY Cepko CL (1990) Lineage-independent determination of cell type in the embryonic mouse retina. Neuron 4:833–845.

Wagner M Thaller C Jessell T Eichele G (1990) Polarizing activity and retinoid synthesis in the floor plate of the neural tube. Nature 345:819–822.

Warf BC Fok-Seang J Miller RH (1991) Evidence for the ventral origin of oligodendrocyte precursors in the rat spinal cord. Cell Diff 11:2477–2488.

Wetts R Fraser SE (1988) Multipotent precursors can give rise to all major cell types of the frog retina. Science 239:1142–1145.

Wetts R Serbedzija GN Fraser E (1989) Cell lineage analysis reveals multipotent precursors in the ciliary margin of the frog retina. Dev Biol 136:254–263.

Williams BP Read J Price J (1991) The generation of neurons and oligodendrocytes from a common precursor cell. Neuron 7:685–693.

Witkowski J (1985) The hunting for the organizer: an episode in biochemical embryology. Trends Biochem Sci 10:379–381.

Wolf MK Brandenberg MC Billings-Cagliardi S (1986) Migration and myelination by adult glial cells: reconstructive analysis of tissue culture experiments. Cell Diff. 6:3731–3738.

Zamora AJ Mutin M (1988) Vimentin and glial fibrillary acidic protein filaments in radial glia of the adult urodele spinal cord. Neuroscience 27:279–288.

The Neuroglia
in the CNS of Teleosts

Juan M. Lara, Almudena Velasco,
José R. Alonso, and José Aijón

1. Introduction

Although the concept of neuroglia, or "nervenkitt" (Virchow, 1846) is prior to the identification in the nervous system of cells distinct from neurons (Deiters, 1865), only once the concept of tissue and cellular theory are accepted can the glia be considered as an integral part of the nervous tissue, differentiated from the neurons, but closely associated with them, both spatially and functionally.

The great conceptual and technical advances in neurohistology at the end of the nineteenth century and beginning of the twentieth century permitted the classification of the different components of the neuroglia of the central nervous system (CNS) of vertebrates by means of specific metallic impregnations for each cell type. In this way not only are the large neuroglial groups described—astrocytes (Deiters, 1865; Ramón y Cajal, 1911), oligodendrocytes (Rio Hortega, 1921; Penfield, 1924), and ependymocytes (Lenhossék, 1891; Achúcarro, 1915)—but also subtypes are determined, with regard to their morphology and distribution, as fibrous and protoplasmic astrocytes (Andriezen, 1893; Ramón y Cajal, 1911) and by their degree of maturity, as radial glia and tanycytes (Ramón y Cajal, 1911; Achúcarro, 1915), by their relationship with other elements of the nervous system, as perineuronal, perivascular or free glia (Ramón y Cajal, 1911), or by their functional state, as reactive astrocytes.

From: *Neuron-Glia Interrelations During Phylogeny: I. Phylogeny and Ontogeny of Glial Cells*
A. Vernadakis and B. Roots, Eds. Humana Press Inc., Totowa, NJ

The techniques of specific impregnation for the distinct glial types of mammals do not give good results in the CNS of bony fish (Ramón y Cajal, 1911). Furthermore, it is not always possible to classify the glial elements as belonging to one of the types defined in mammals (Kruger and Maxwell, 1967; Lara et al., 1989). The electron microscopy confirms the existence of glial cells of diverse characteristics in the CNS of teleosts that, in the majority, fit, with some variations, the descriptions of glial cells in mammals (Kruger and Maxwell, 1966, 1967). There are, however, an abundance of glial elements with characteristics intermediate between different types (Lara, 1982; Lara et al., 1989). The existence of these intermediate elements between varieties of one single glial type in higher vertebrates has been known since the last century (Ramón y Cajal, 1890) and more recently transitions between different glial types have been described (Reyners et al., 1982). Ramón y Cajal (1890) considered these transitional forms as morphological transformations, whereas in more recent works they are considered as exponents of different degrees of differentiation of multipotential cells (Vaugh and Peters, 1968; Vaugh et al., 1970). In teleosts this morphogenetic transformation or expression of multipotentiality seems to be particularly wide (Lara et al., 1989; Velasco, 1992).

In this work we analyze three parts of the CNS of teleosts (optic nerve, olfactory bulb and optic tectum), with a very different neuroglial organization

2. Methodology for the Study of the Neuroglia in Teleosts

The routine morphological techniques offer little information about the glia of teleosts: In general, they only permit the differentiation between neuronal and nonneuronal elements. The use of semi-thin sections (1–2 μm) of tissue embedded in plastic or resin offers greater information, but is insufficient in the majority of cases for the unequivocal identification of the different glial types (Lara et al., 1987, 1989; Velasco, 1992).

Some variants of Golgi's technique occasionally offer images of glial elements. The glial cells most frequently impregnated are the telencephalic and tectal ependymocytes (Vanegas et al., 1974; Stevenson and Yoon, 1982; Miguel et al., 1986; Lara et al., 1989;

Velasco, 1992), although it is also possible to identify astrocytes in any encephalic structure (Velasco, 1992), with the exception of the optic tectum. With only rare exceptions, the analysis of these images using the traditional techniques of light microscopy offers little data about glial morphology; the use of confocal microscopy partially solves this problem (Velasco, 1992).

The utility of light microscopy in the study of the neuroglia of teleosts has increased with the application of immunocytochemical techniques. As in the other vertebrates, the localization of antigens of intermediate filaments of glial cells, such as GFAP or vimentin (Levine, 1989, 1993; Nona et al., 1989, 1992; Cardone and Roots, 1990; Stafford et al., 1990; Blaugrund et al., 1991; Cohen et al., 1993) or of proteins related to glial metabolism, like S-100 (Cohen et al., 1993) are indispensable tools for the study of the astroglia and ependymal cells; whereas the detection of specific enzymes, such as 2',3'-cyclic nucleotide 3'-phosphodiesterase (CNP) (Sprinkle, 1989), galactocerebroside (Ransch et al., 1982), and myelin specific proteins (Jeserich and Wachneldt, 1986, 1987; Waehneldt et al., 1986; Jeserich et al., 1990; Jeserich and Stratmann, 1992) are useful for the detection of oligodendrocytes. However, after 200 million years of divergent evolution, in interpretation of the results of these techniques special care is recommended.

The techniques of conventional electron microscopy, alone or combined with silver impregnations or immunocytochemical marking, constitute another fundamental methodology for glial analysis in teleosts (Kruger and Maxwell, 1966, 1967; Wolburg, 1978; Kosaka and Hama, 1983; Wolburg and Bouzehouane, 1986; Lara et al., 1989; Nona et al., 1992).

3. Cell Types

As in the rest of the vertebrates, the glial population of the CNS of teleosts is much greater than the neuronal population. An important part of these nonneuronal cells belongs to the controversial microglia whose ontogenetic and functional relationships with the other elements of the central glia are not well established (Fedoroff and Hao, 1991). These aspects acquire great relevance in lower vertebrates because of the particular relationships of the neuroglia and microglia in establishing the environment appropriate for the regeneration of the CNS (Dowling et al., 1991).

With regard to the remaining glial population, three basic types of neuroglia are present in the CNS of all the vertebrates: ependymocytes, astrocytes, and oligodendrocytes. In mammals, their identification and classification according to their morphological and immunocytochemical characteristics are well established (Bignami et al., 1980; Sommer and Schachner, 1981; Dahl et al., 1986; Privat and Rotaboul, 1986; Miller et al., 1989; Sprinkle, 1989; Schmidt-Kastner et al., 1993; Ehinger et al., 1994). These characteristics serve as the basis for the identification of the distinct types of neuroglia in the rest of the vertebrates (Kruger and Maxwell, 1967), however, the evolutive separation is reflected structurally and functionally in the glial cells of the teleosts. Moreover, the morphological and immunocytochemical characteristics, the distribution and, possibly, the function of these cells are not homogeneous in the different zones of the central nervous tissue of this class of vertebrates.

3.1. Ependymocytes

The ependymocytes are the principal neuroglial type in lower vertebrates (Studnicka, 1900; Ramón y Cajal, 1911; Kruger and Maxwell, 1967) and carry out some of the functions typical of astrocytes in mammals (Ramón y Cajal, 1911; Achúcarro, 1915; Kruger and Maxwell, 1967). The reduction in the number of ependymal cells in favor of other glial types seems to be associated with the evolutive process in vertebrates (Roots, 1986). This neuroglial type consists of cells of neuroectodermic origin of early ontogenetic development that initially cross the whole thickness of the neural tube, connecting the ventricular cavities with the pial surface. In some zones of the CNS of teleosts this arrangement endures in adult animals (Kruger and Maxwell, 1966, 1967; Vanegas et al., 1974; Stevenson and Yoon, 1982; Lara et al., 1989). In the ventricular zone the somata of these cells constitute a continuous and closed weft of monostratified structure or, in the majority of cases, pseudostratified, which project cilia and microvilli toward the ventricular cavity (Kruger and Maxwell, 1967; Roots, 1986; Alonso, 1987; Lara et al., 1989). From the cell body a radial prolongation originates which enters the nervous tissue and, when it reaches the subpial zones, participates in the formation of the subpial limitant (Kruger and Maxwell, 1967; Stevenson and Yoon, 1982; Roots, 1986; Lara et al., 1989). In teleosts, the morphology, ultrastructure, and immunocytochemical characteristics of the ependymocytes are not homogeneous throughout the whole extension of the CNS.

3.1.1. Ependymocytes in the Olfactory Bulb

Ependymocytes are present in the olfactory bulb of all the vertebrates, although they are more evident in lower vertebrates (Garrido, 1978). In this structure of teleosts the ependymocytes are located surrounding the ventricle and are present in both the olfactory bulb and tract (Fig. 1). Their arrangement is pseudostratified and they present diverse morphologies, easily observed using silver impregnations (Alonso, 1987). In each of these cells we can consider three parts: a cell body, a "neck" that connects it with the ventricular surface, and a radial expansion that enters the bulb structure (Fig. 2A).

The cell body is the portion of greatest diameter and its form varies from flattened to cubic or high prismatic shape.

The neck can be of very varied size. In Cyprinodontes it is usually very short, whereas in Salmonidae it can be quite long, and in Cyprinidae it may be nonexistent or extremely short for some ependymocytes, although in others the neck can be relatively long.

The radial expansion of the cell body in some cells does not divide, whereas in others there is a bifurcated opening (Fig. 1). The surface of the expansions can be smooth, as in some cells of Cyprinidae, or densely covered with numerous short expansions, as in Salmonidae (Alonso, 1987).

Electron microscopy of the ependymocytes shows a voluminous nucleus in relation to the size of the cell body, with finely dispersed chromatin, and some peripheral accumulations (Fig. 2B). The dictyosomes and a good number of mitochondria are usually located in the base zones of the cell body; the smooth endoplasmic reticulum is quite developed, whereas the rough cisternae and microtubules are scarce. Surrounding the nucleus and in the radial expansion bundles of intermediate filaments are located parallel to the principal axis of the cell. Other filaments are dispersed in the cell body and the "neck," these become denser in the cell periphery. Moreover, the radial expansion contains some cisternae of smooth endoplasmic reticulum, mitochondria, and microtubules arranged in parallel (Alonso, 1987).

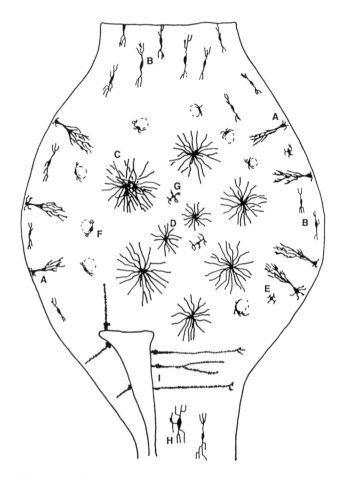

Fig. 1. Scheme of the neuroglial cytoarchitecure of the olfactory bulb of teleosts obtained using silver impregnations. **(A)** Astrocytes from the subpial limitant. **(B)** Astrocytes of the olfactory nerve and the olfactory nerve fiber layer. **(C)** Astrocytes of the plexiform layer. **(D)** Astrocytes of the granule cell layer. **(E)** Oligodendrocytes of the glomerular layer. **(F)** Perineuronal satellite Oligodendrocytes. **(G)** Oligodendrocytes of the granule cell layer. **(H)** Oligodendrocytes of the olfactory tract. **(I)** Ependymocytes.

The portions of the ependymocyte in contact with the ventricular cavity show cilia and irregular microvilli-like expansions (Fig. 2B). A great part of the plasma membrane of the soma and the "neck" of adjacent ependymocytes is occupied by interdigitations, adherent zonulae, desmosomes, and tight junctions (Fig. 2B).

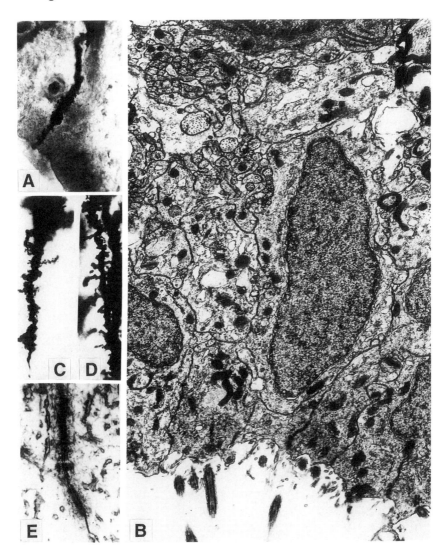

Fig. 2. Ependymocytes in the CNS of teleosts. **(A)** Ependymocyte of the olfactory bulb of *Oncorhynchus mikiss*. Golgi Kopsch. **(B)** Cell body, neck and ventricular projections of ependymocytes in the olfactory bulb of *Gobio gobio*. **(C)** Ependymocyte of the optic tectum of *Cyprinus carpio*. Golgi- Colonnier. (From Lara et al., 1989.) **(D)** Lateral branches of the ascendent prolongation of an ependymocyte of *Cyprinus carpio*. Golgi Colonnier. (From Lara et al., 1989.) **(E)** Junction complex between the lateral areas of ependymocyte perikarya in the optic tectum of *Cyprynus carpio*. (From Lara et al., 1989.)

3.1.2. Ependymocytes in the Optic Tectum

In the optic tectum of teleosts ependymocytes constitute the only glial type constantly observable using silver impregnations (Ramón y Cajal, 1911; Kruger and Maxwell, 1967; Vanegas et al., 1974; Miguel et al., 1986; Lara et al., 1989).

The ependymal somata are situated in the deeper zone of the periventricular stratum (SPV); as in the olfactory bulb, ependymocytes with soma adjacent to the ventricle are found and others with the cell body somewhat distanced from the ventricular surface, but with a "neck" that reaches this surface. At least in Cyprinidae both types of tectal ependymocytes coexist in the same species (Miguel et al., 1986; Lara et al., 1989). The radial prolongation crosses all the tectal strata until it reaches the pial surface (Fig. 2C); throughout its course it emits numerous short fine prolongations (Fig. 2D) (Vanegas et al., 1974; Miguel et al., 1986; Lara et al., 1989).

The ultrastructural characteristics of these glial cells have been described in diverse species of teleosts (Kruger and Maxwell, 1966, 1967; Laufer and Vanegas, 1974; Roots, 1986; Stevenson and Yoon, 1982; Lara et al., 1989) and are similar to those of their homologs in the olfactory bulb, including the abundance of interdigitations, adherent zonulae, desmosomes, and tight junctions near the ventricular limit (Fig. 2E). The intermediate filaments organized in bundles are, in general, less abundant in the somata of the tectal ependymocytes than in those of the olfactory bulb. In the radial prolongation, these bundles of filaments are relatively abundant near the soma. In the subpial expansions the filaments are also abundant, but the arrangement in loose extensions predominates over the organization in bundles. Moreover, the bundles of filaments are more evident in Cyprinidae (Roots, 1986) than in Salmonidae.

Although in higher vertebrates GFAP is considered a specific marker for astrocytes (Dahl and Bignami, 1973; Hájos and Kálmán, 1989; Schmidt-Kastner and Szymas, 1990), in adult animals of diverse species of teleosts the ependymal glia contain GFAP (Anderson et al., 1984: Onteniente et al., 1983; Cardone and Roots, 1990; Velasco, 1992), and in *Tinca tinca* and *Oncorhynchus mykiss* the immunostaining is more intense in periventricular zones (in the somata and origin of the radial expansions) and in the subpials (in the endfeet that constitute the subpial limitant). Moreover, vimentin, considered to be the major component in the cytoskeleton of immature astroglial cells (Dahl et al., 1981; Rickmann et al., 1987) and of specific astrocytes (Osborn et al., 1984; Schmidt-Kastner and Szymas, 1990), also causes a similar immunostaining to GFAP in the optic tectum (Cardone and Roots, 1990; Velasco, 1992). These two proteins with intermediate filaments can also be considered to constitute individualized filaments or to coexist in the same filaments (Cardone and Roots, 1990).

3.2. Astrocytes

The astrocytes of teleosts, like those of the rest of the vertebrates, are non-neuronal cells of neuroectodermic origin morphologically characterized by a cell body of irregular shape from which numerous relatively long processes originate. Some of these make contact with blood vessels, others participate in the structure of the subpial limitant, and others are involved in the isolation of synapses both in individuals and groups (Lara and Aijón, 1983a; Roots, 1986). The most notable characteristic of this cellular type in all vertebrates is the organization and composition of its cytoskeleton, comprised of gliofibrils, ultrastructurally and immunocytochemically identified as intermediate filaments (Trimmer et al., 1982; Vaughan, 1984; Dahl et al., 1986; Goetschy et al., 1986; Kalnins et al., 1986).

The aforementioned inconstancy of silver impregnation for these cells in the CNS of teleosts considerably reduces the possibility of accurate identification. Thus the absence of impregnated astrocytes in zones like the optic tectum has led some authors to consider that they do not exist in this structure (Ramón y Cajal, 1911; Vanegas et al., 1974). However, electron microscopy shows that in all the zones of the CNS of these animals glial cells with characteristics like those of astrocytes are found: a relatively euchromatic nucleus with peripheral chromatic accumulations, surrounded by a generally low density cytoplasm, with few organules and with some dense bodies (Kruger and Maxwell, 1967; Alonso, 1987; Lara et al., 1989). Notable differences with regard to mammal astrocytes are that these are more rich in glycogen and have fewer gliofilaments (Kruger and Maxwell, 1967; Roots, 1986) that, what is more, are usually particularly scarce in the distal zones of the prolongations.

3.2.1. Astrocytes in the Optic Nerve

In normal conditions the optic nerve of teleosts is a folded ribbon comprised of axons of ganglion cells of the retina. Almost all are myelinized (Wolburg, 1978, 1981; Easter et al., 1981,1984), highly organized (Rusoff and Easter, 1980; Easter et al., 1981, 1984; Magss and Scholes, 1990), and arranged in fascicles (Maggs and Scholes, 1990). Two types of neuroglial cells participate in its structure: oligodendrocytes and astrocytes. The oligodendrocytes form the myelinic sheath, whereas the astrocytes intervene in the organization of fascicles and contribute to the flexibility and mechanical stability of the nerve (Maggs and Scholes, 1990).

The astrocytes of the optic nerve of teleosts, even though they fit into the above mentioned pattern, have particular characteristics that differentiate them both from the astrocytes of the optic nerve of mammals and the other astrocytes of the teleost CNS. In the optic nerve of teleosts, only one type of astrocyte is considered, that denominated reticular (Maggs and Scholes, 1990), morphologically characterized by their more or less regular distribution along the nerve and the organization of their prolongations that participate in the bundling of the nerve, whereas through desmosomes (Skoff et al., 1986), tight, and gap junctions they mechanically stabilize the nerve and confer flexibility (Maggs and Scholes, 1990).

Immunocytochemically, the particularities of these astrocytes are even more notable, although in nerves without lesions these cells possess glial fibrillary acidic protein (GFAP) (Markl and Frank, 1988; Levine, 1991; Cohen et al., 1993). Its immunohistochemical detection is very low or nonexistent (Quitschke and Schecher, 1984; Levine, 1989: Nona et al., 1989; Cardone and Roots, 1990; Blaugrund et al., 1991); on the other hand, lesion of the optic nerve causes strong immunoreactivity for GFAP in astrocytes near the zone of lesion, although in the zone itself GFAP is not expressed (Stafford et al., 1990; Cohen and Schwartz, 1993). Furthermore, the immunoreactivity for vimentin also disappears from the site of the lesion immediately when this is produced, but reappears in relationship to the regeneration of the axons (Blaugrund et al., 1992; Cohen et al., 1993).

3.2.2. Astrocytes in the Olfactory Bulb

In teleost olfactory bulb, the astrocytes comprise the most numerous glial population. Silver impregnations show morphological variations of the astrocytes according to their location (Fig. 3A) (Alonso, 1987).

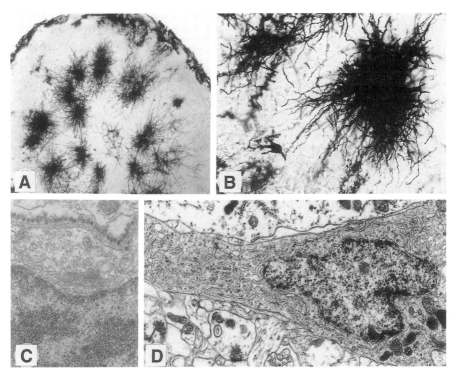

Fig. 3. Astrocytes in the CNS of teleosts. **(A)** Groups of astrocytes in the plexiform and granule cell layers of the olfactory bulb of *Cyprinus carpio*. Golgi-Colonnier. **(B)** Astrocytes in the plexiform layer of Barbus bocagei. **(C)** Astrocyte cell body near subpial limitant of the olfactory bulb of *Cyprinus carpio*. (From Lara et al., 1987). **(D)** Astrocyte-like cell perineuronal satellite in the optic tectum of *Cyprinus carpio*. (From Lara et al., 1989).

In the peripheral parts of the olfactory nerve layer, the astrocytes usually have a pyriform soma from which a long prolongation with many ramifications and centripetal orientation projects and four to five prolongations without ramifications, parallel to the surface, that terminate in bulbous ends (Fig. 1). More internally, between the packets of fibers, the astrocytes have fusiform somata and prolongations parallel to the bundles that they sheathe (Fig. 1). In mammals, differences have been described in the density in the cytoplasm matrix of the astrocytes of this layer (Berger, 1971; Doucette, 1984) and in the filament content, which leads to the consideration of two different types of astrocytes, the typical ones and the "ensheathing cells," that may have different functions (Doucette, 1984, 1986). In teleosts, the two astroglial types of this stratum are differentiated in the quantity of heterochromatin of their nuclei, but the quantity of filaments of their cytoplasm is similar. The axon packets are surrounded by prolongations of very low density, originating in the astrocytes with clearer nuclei, whereas the more peripheral ones, with more heterochromatic nuclei, do not appear to fulfill this function. A great number of these latter astrocytes participate in the formation of subpial glial limitant (Fig. 3C) (Lara et al., 1987), and it is not rare to observe mitotic figures in these cells (Alonso et al., 1989a).

In the glomerular layer, the astrocytes are more scarce than the oligodendrocytes. Their morphology is varied depending on their situation in the glomerular zone or in the nest (Alonso, 1987), although the peculiar structure of this stratum in bony fish (Alonso et al., 1989b) does not permit the differentiation between glomerular and periglomerular astrocytes found in higher vertebrates (Gayoso et al., 1977).

In the plexiform layer the astrocytes are large (Fig. 3B) and on occasion, appear in small groups (Fig. 1). They emit long prolongations, without ramifications, in all directions, which gives them a star-like appearance.

In the granular layer the astrocytes have a similar morphology to those of the plexiform layer, although they are smaller and more scarce (Fig. 1).

In the subependymal layer the astrocytes show fusiform somata and prolongations parallel to the ventricular surface.

3.2.3. Astrocytes in the Optic Tectum

Some authors exclude the astrocytes from the glial types present in the optic tectum of teleosts (Ramón y Cajal, 1911; Laufer and Vanegas, 1974), or their number is considered to be very low (Wolburg and Bouzehouane, 1986), and their habitual functions to be carried out by ependymocytes or variants of these (Stevenson and Yoon, 1982). Silver impregnations seem to confirm this scarcity of astrocytes in the tectum mesencephali, since astrocyte-like cells are impregnated only on a number of occasions, and in an irregular way (Lara et al., 1989), and generally in periventricular zones. They are usually periventricular elements with irregular cell bodies and with prolongations with little ramification.

The ultrastructure of these cells is that which is typical of astrocytes of lower vertebrates (Kruger and Maxwell, 1967), even though the gliofilaments are particularly scarce and moreover are drastically reduced on the prolongations as these become more distant from the soma, thus it is not surprising that their immunoreactivity for GFAP and vimentin is very low, with notable variations from one species to another (Velasco, 1992).

Besides these perivascular astrocyte-like cells, electron microscopy shows the presence of somata with astroglial characteristics in the whole tectal thickness. In these internal strata, the somata of the astrocyte-like cells are usually closely associated with the neuronal perikarya (Fig. 3D). Their prolongations occasionally make contact with each other through gap junctions and, in the subpial zones, tight junctions are frequently found between astrocyte-like prolongations and terminal processes of ependymocytes (Lara et al., 1989).

In the optic tectum of teleosts, cells with intermediate characteristics between astrocytes and oligodendrocytes are particularly frequently found (Lara et al., 1989), not only in the periventricular stratum, the principal zone of cellular proliferation and differentiation, but also in the other tectal strata. We have only considered the elements with characteristics typical of well-defined astrocytes as astrocytes.

3.3. Oligodendrocytes

Rio Hortega (1921) denominated glial cells characterized by their scarce radiations of oligodendrocytes. This glial type is formed by cells of neuroectodermic origin whose development from a precursor common to type 2 astrocytes, denominated 0-2A , seems to be general in all the vertebrates (Raff et al., 1983; Jeserich and Stratmann, 1992; Sivron et al., 1992; Vauyiouklis and Brophy, 1993).

The techniques of metal impregnation do not yield good results for this glial type in lower vertebrates (Kruger and Maxwell, 1967; Lara et al., 1989). Only histochemical and immunocytochemical methods demonstrate unequivocally their presence in the CNS of teleosts under the light microscope (Jeserich, 1981; Jeserich et al., 1990; Sivron et al., 1991; Jeserich and Stratmann, 1992; Nona et al., 1992).

Ultrastructurally, those glial cells of teleosts with characteristics comparable to those of this cell type in mammals are identified as oligodendrocytes, typically. The nucleus is sufficiently more heterochromatic than that of the astrocytes, with electrodense cytoplasm with some cisternae of the RER, numerous free ribosomes, and an extense system of microtubules, the gliofibrils, and the grains of glycogen being absent (Kruger and Maxwell, 1967; Alonso, 1987). They do not participate directly in the formation of the hematoencephalic barrier and are associated with the myelinic sheaths (Kruger and Maxwell, 1967).

3.3.1. Oligodendrocytes in the Optic Nerve

The oligodendrocytes of the optic nerve are the origin of the myelinic sheaths that affect almost all the axons which pass through the optic nerve. The constant growth of the retina of the teleosts causes the constant addition of new axons originating from the new ganglion cells of the peripheral retina (Rusoff and Easter, 1980; Easter et al., 1981, 1984), which determines that in the folded ribbon that constitutes the nerve, as well as in the retinotopic organization a chronological order exists that is evident in the recently incorporated axons through their progressive myelinization (Wolburg, 1981; Maggs and Scholes, 1990). In undamaged nerves of *Cyprinus carpio* and *Onchorynchus mykiss*, the distribution of oligodendrocytes in the anterior-posterior plane offers a sufficiently homogeneous appearance (Caminos, 1993), which leads to the supposition that the number of oligodendrocytes of this structure varies in proportion to the length of the nerve and the number of axons that comprise it, parameters determined for the species analyzed, and the age of the animal.

Ultrastructurally, the oligodendrocytes of the optic nerve of teleosts present the typical characteristics of this glial type in teleosts, the elevated density of their cytoplasm being emphasized, with relatively thick expansions that enter between the little-myelinized axon packets; in proportion to the development of the myelinic sheaths, the oligodendroglial prolongations become less evident and in the mature zones of the optic nerve only the perikaryon and the commencement of some expansions together with the myelinic sheaths are usually identifiable.

Although some data are available for the development of these oligodendrocytes from type O-2A cells (Jeserich et al., 1990;

Sivron et al., 1992; Jeserich and Stratmann, 1992; Lotan and Schwartz, 1992), the majority of recent studies are concerned with the processes of myelinization during the regeneration of the optic nerve (Wolburg, 1981; Jeserich et al., 1990; Nona et al., 1992). The particular glial modifications in the zone of the lesion (Stafford et al., 1990; Levine 1991; Nona et al., 1992; Fuchs et al., 1994) have led to the proposition that in the zone of lesion it is not central neuroglia but Schwann cells that are responsible for the myelinization of the axons in regeneration (Nona et al., 1992).

3.3.2. Oligodendrocytes in the Olfactory Bulb

Silver impregnations only show some oligodendrocytes with rounded or fusiform cell bodies from that originate a small number of prolongations which occasionally bifurcate (Alonso, 1987). On electron microscopy, the most notable characteristic of these oligodendrocytes is the wide variability of the cytoplasm density— from somewhat more dense than the typical astrocytes to very dark cells.

Oligodendrocytes are located in all the zones of the olfactory bulb in teleosts (Fig. 1). They are especially abundant in the granular layer, but they are also found as interfascicular oligodendrocytes in the layer of the fibers of the olfactory nerve, related to the myelinic sheaths of the caudal zone of the bulb, or as perineuronal satellites of mitral and ruffed cells.

The high number of oligodendrocytes in the granular layer is also observed in higher vertebrates (Willey, 1973), but in teleosts these cells are isolated and associations between their prolongations is not observed. Between the perinuclear satellites and the neurons with which they associate, relationships distinct from simple contiguity with their plasma membranes have not been located, either.

With regard to the association with myelinic sheaths, the continuity of the oligodendroglial cytoplasm with myelinic sheaths of axons is frequently observed, but, what is more, they are also found associated with neuronal somata and myelinated dendrites (Lara and Aijón, 1983b; Alonso, 1987).

In the olfactory bulb of higher vertebrates myelinated neuronal somata with interspecific differences of distribution have been described (Willey, 1973; Braak et al., 1977; Díaz-Flores et al., 1977; Burd, 1980; Tigges and Tigges, 1980). In teleosts, the myelinated somata are very scarce in the olfactory bulb, but in all the species studied some are located in the plexiform layer (Alonso, 1987). They are particularly abundant in *Cyprinus carpio*, especially in animals of great size and of age superior to 4 yr.

In teleosts, these myelinated somata can belong to three neuronal types: short axon neurons, star-shaped neurons, and granules. The myelinic sheath is usually thin, of four to six layers (Fig. 4A), and on occasion, its association with oligodendrocytes can be observed (Lara and Aijón, 1983b; Alonso, 1987).

The myelinated dendrites in the olfactory bulb have been described in diverse species of mammals (Pinching, 1971; Willey, 1973; Braak et al., 1977; Burd, 1980; Tigges and Tigges, 1980; Rehn et al., 1986). In teleosts, although they are infrequent in the olfactory bulb, myelinated dendritic segments have been described in various species of the Cyprinidae and Salmonidae families (Alonso, 1987). They are thick dendrites that can be totally or partially encased by a myelin sheath of 5–10 layers (Fig. 4B). They are located in the glomerular and plexiform layers, and are particularly frequent in the limit between these strata. Whereas in mammals the myelinated dendrites can originate in mitral, tufted, or periglomerular neurons (Pinching, 1971; Pinching and Powell, 1971; Willey, 1973; Tigges and Tigges, 1980; Rehn et al., 1986), in teleosts they all originate from mitral cells, since morphologically and ultrastructurally they adjust to the characteristics of the dendrites of these neurons and, moreover, neither tufted nor periglomerular neurons exist (Alonso, 1987; Alonso et al., 1989b).

3.3.3. Oligodendrocytes in the Optic Tectum

In this structure the electron microscope shows a wide distribution of glial cells with characteristics of oligodendrocytes (Laufer and Vanegas, 1974; Lara et al., 1989). As in the olfactory bulb, these cells can occupy three different arrangements: oligodendrocytes dispersed throughout the neuropil , associated with zones rich in myelin and perineuronal satellites (Lara et al., 1989).

In all the cases, the ultrastructural characteristics are the basic ones of the oligodendrocytes of lower vertebrates (Kruger and Maxwell, 1967), even if in the cells with less electrodense cytoplasm the microtubules are scarce, the dense bodies are frequent and, on some occasions, some filaments are distinguished, thus they could be cataloged ultrastructurally as intermediate elements between astrocytes and oligodendrocytes. Between these elements and the cells with unequivocal characteristics of oligodendrocytes there is an extensive variety of intermediate elements (Lara et al., 1989).

In the stratum periventriculare (SPV) cells unequivocally identified as oligodendrocytes are scarce. The oligodendrocytes in the stratum marginale (SM) are even more scarce. On the other hand, they are abundant in the strata rich in myelin: stratum album centrale (SAC) and stratum opticum (SO) (Fig. 4C) (Laufer and Vanegas, 1974; Jeserich, 1981; Lara, 1982). In the central strata of the optic tectum, stratum griseum centrale (SGC) and stratum fibrosum et griseum superficiale (SFGS), there is a certain gradation from the deeper to the upper zones with regard to the proportion of clear and dark oligodendrocytes: The former are more abundant than the latter in the deeper zones with this relationship being inverted in the upper zones.

With regard to the oligodendrocytes, perineuronal satellites are abundant in the SFGS and are frequently associated with pyriform neurons (Fig. 4D) (Lara and Aijón, 1983a; Lara et al., 1989); as in the olfactory bulb, their morphological relationship with the neurons with which they associate is only of contiguity.

The majority if not the totality of the oligodendrocytes of the SAC and SO stata are associated with large myelinic packets; the oligodendrocytes of the neuropil of the SGC and SFGS strata are usually in the proximities of myelinized axons (Fig. 4E). On the other hand, the associations of perineuronal satellite oligodendrocytes with myelinic structures are not frequent.

4. Comparative Aspects of the Teleost Glial Typology

In accordance with the traditional classification of the neuroglia, ependymocytes, astrocytes, and oligodendrocytes have been described in the CNS of teleosts, although with some particularities that differentiate them from the neuroglial types of higher vertebrates. Moreover, in each of the three structures analyzed, the optic nerve, the olfactory bulb and the optic tectum, the neuroglia possess some particular characteristic.

The ependymocytes, both in the olfactory bulb and in the optic tectum present morphological and ultrastructural charac-

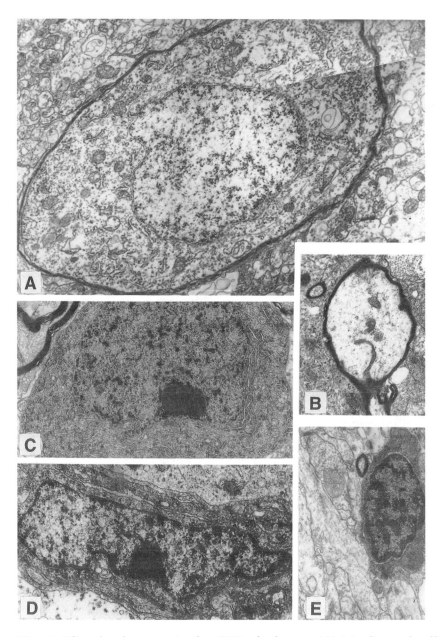

Fig. 4. Oligodendrocytes in the CNS of teleosts. **(A)** Myelinated cell body of the plexiform layer of the olfactory bulb of *Cyprinus carpio*. **(B)** Myelinated dendrite in the glomerular layer of the olfactory bulb of *Cyprinus carpio*. **(C)** Cell body of a dense oligodendrocyte in the optic tectum of *Barbus bocagei*. (From Lara et al., 1989). **(D)** Oligodendrocyte

teristics sufficiently similar to each other and comparable with the basic characteristics of the ependymal cells of other vertebrates (Kruger and Maxwell, 1966, 1967; Laufer and Vanegas, 1974, Stevenson and Yoon, 1982; Roots, 1986). However, whereas in the olfactory bulb the ependymocytes comprise a relatively scarce population in relation to the total number of demonstrable neuroglial cells, in the optic tectum they constitute by far the best represented neuroglial type.

The ependymal cells of the optic tectum are also the most studied. Although ependymocytes with and without "neck," with a single expansion, and with expansions with ramifications, smooth and with numerous small expansions have been described (Lara, 1982; Miguel et al., 1986; Alonso, 1987; Velasco, 1992), all of them are considered to be variants of one unique glial type and coincide basically with the descriptions of this glial type in diverse species of teleosts (Kruger and Maxwell, 1966, 1967; Laufer and Vanegas, 1974; Onteniente et al., 1983; Roots, 1986), although on rare occasions packets of gliofilaments as dense as those described in tectal ependymocytes of goldfish (Roots, 1986), or the organization of planar tubules of the marine sandbass (Kruger and Maxwell, 1966) have been observed. Nevertheless, some authors consider two cell types with distinct morphology and ultrastructure (Stevenson and Yoon, 1982).

With regard to the astrocytes, those of the optic nerve can be considered distinct from the other astrocytes of the CNS of teleosts because of their particular spatial distribution, the frequency and organization of their systems of anchorage, in particular desmosomes (Maggs and Scholes, 1990), their immunocytochemical characteristics (Nona et al., 1989; Cardone and Roots, 1990; Blaugrund et al., 1991; Cohen and Schwartz, 1993) and their modifications during the regenerative processes (Braverman et al., 1989; Levine, 1991, 1993; Sivron et al., 1991; Cohen et al., 1993; Fuchs et al. 1994). Regarding the astrocytes of the olfactory bulb, although they conserve the general characteristic of this glial type in teleosts of possessing very scarce gliofilaments, because of their morphology, ultrastructure, and immunocytochemical characteristics,

---

satellite perineuronal in the optic tectum of *Cyprinus carpio*. (From Lara et al., 1989). **(E)** Oligodendrocyte associated with a thin myelinic axon in the tectum neuropil of *Barbus bocagei*.

they can be considered as the most similar to the astrocytes of mammals. In the optic tectum, on the other hand, even the existence of this glial type is debated. The inefficiency of the classical methods of specific staining, the abundance of ependymocytes that could carry out a considerable part of the functions of the astrocytes and the existence of a great number of cells with intermediate ultrastructural characteristics between astrocytes and oligodendrocytes explain the reticence in recognizing the existence of tectal astrocytes. What is more, the presence in ependymocytes of characteristic markers of astrocytes could disguise the immunostaining of these latter cells, which are much more scarce.

The oligodendrocytes of teleosts possibly all originate from type O-2A cells (Jeserich and Stratmann, 1992; Sivron et al., 1992) and follow similar maturing processes (Jeserich, 1981; Jeserich et al., 1990), forming myelinic sheaths as a principal function. However, the oligodendrocytes of the optic nerve, the most homogeneous of those studied in this work, are particularly permissive in terms of the regenerative processes (Blastmeyer et al., 1990; Sivron et al., 1991) and it is even possible that during the regeneration they are substituted in the zone of the lesion by glia typical of the PNS (Nona et al., 1992). The oligodendrocytes of the olfactory bulb are characterized by the infrequent capacity to form myelinic sheaths that affect the somata and dendrites. The tectal oligodendrocytes are a heterogeneous family with a diversity of ultrastructural characteristics that could be interpreted as different grades of maturity of a single glial type or as a transition between two different glial types, with neither possibility being exclusive.

5. Concluding Remarks

The classification of the neuroglia used in higher vertebrates is applicable, along general lines, to the teleosts, but with morphological, ultrastructural, immunocytochemical, and functional specifications. Furthermore, the neuroglial types present in each one of the three studied parts of the CNS, the optic nerve, the olfactory bulb, and the optic tectum, possess some characteristics that permit their inclusion in some of the traditional neuroglial types, together with particularities that could correspond to evolutive and/or functional adaptations of the teleosts in general and of each nerve center and glial type in particular.

Acknowledgments

The authors are greatly indebted to GH Jenkins for revising the English style of the manuscript. This work was supported by the DGICyT (PB91-0424).

References

Achúcarro N (1915) De l' évolution de la neuroglie et specialement de ses relations avec l'appareil vasculaire. Trab Lab Invest Biol 13:169–212.

Alonso JR (1987) El bulbo olfatorio de teleósteos de agua dulce: Estructura y ultarestructura. Doctoral Thesis. Univ Salamanca.

Alonso JR Lara J Vecino E Coveñas R Aijón J (1989a) Cell proliferation in the olfactory bulb of adult freshwater teleosts. J Anat 163:155–163.

Alonso JR Piñuela C Vecino E Coveñas R Lara J Aijón J (1989b) Comparative study of the anatomy and laminar organization in the olfactory bulb of three orders of freshwater teleosts. Gegenbaurs morphol Jahrb 135: 241–254.

Anderson MJ Swanson KA Waxman SG Eng LF (1984) Glial fibrillary acidic protein in regenerating teleost spinal cord. J Histochem Cytochem 32:1099–1106.

Andriezen WL (1893) The neuroglia elements in the human brain. Br Med J 2:227–230

Berger B (1971) Etude ultrastructurale de la degenerescence wallerienne experimentale d'un nerf entierement amyelinique: II. Reactions cellulaires. J Ultrastruct Res 37:479–494.

Bignami A Dahl D Rueger DC (1980) Glial fibrillary acidic protein (GFAP) in normal neural cells and in pathological conditions. Adv Cell Neurobiol 1:285–319.

Blastmeyer M Vielmetter J Jeserich G Stuermer CAO (1990) Growth of retinal axons on goldfish optic nerve oligodendrocytes in vitro. Soc. Neurosci Abstr 16:1600.

Blaugrund E Cohen I Shani Y Schwartz M (1991) Glial fibrillary acidic protein in the fish optic nerve. Glia 4:393–399.

Blaugrund E Lavie V Cohen I Schreyer DJ Schwartz M (1992) Axonal regeneration is associated with glial migration: Comparison between the injured optic nerve of fish and rats. J Comp Neurol 330:105–112.

Braak E Braak H Strenge H (1977) The fine structure of myelinated nerve cell bodies in the bulbus olfactorius of man. Cell Tissue Res 182:221–233.

Braverman SB Rappaport I Sharma SC (1989) Characterization of a goldfish antigen during development and regeneration of the visual system. Visual Neurosci 2:449–454.

Burd GD (1980) Myelinated dendrites and neuronal perikarya in the olfactory bulb of the mouse. Brain Res 181:450–454.

Caminos E (1993) Efecto de la temperatura sobre el transporte axónico en la tenca *Tinca tinca*. Estudio con trazadores neuronales. Minor Thesis. Univ Salamanca.

Cardone B Roots BI (1990) Comparative immunohistochemical study of glial filament proteins (glial fibrillary acidic protein and vimentin) in goldfish octopus and snail. Glia 3:180–192.

Cohen I Schwartz M (1993) cDNA clones from fish optic nerve. J Comp Biochem Physiol 104:439–447.

Cohen I Shani Y Schwartz M (1993) Cloning and characteristics of fish glial fibrillary acidic protein: Implications for optic nerve regeneration. J Comp Neurol 334:431–443.

Dahl D Bignami A (1973) Immunochemical and immunofluorescence studies of the glial fibrillary acidic protein in vertebrates. Brain Res 61:279–293.

Dahl D Björklund H Bignami A (1986) Immunological markers in astrocytes, in Astrocytes, vol. 3 (Fedoroff S Vernadakis A, eds.), Academic, Orlando, FL pp. 1–25.

Dahl D Reguer DC Bignami A Weber K Osborn M (1981) Vimentin, the 57,000 molecular weight protein of fibroblast filaments is the major cytoskeletal component in immature glia. Eur J Cell Biol 24:191–196.

Deiters O (1865) Untersuchungen über Gehirn und Rückenmark des Menschen und der Säugethiere (Schulze M, ed.), Braunschweig.

Díaz-Flores L Gayoso MJ Garrido M (1977) Origin of the myelin sheath in myelinated nerve cell bodies of the olfactory bulb. Morf Norm Patol Sec A 1:339–342.

Doucette JR (1984) The glial cells in the nerve fiber layer of the rat olfactory bulb. Anat Rec 210:385–391.

Doucette JR (1986) Astrocytes in the olfactory bulb, in Astrocytes, vol. 1, (Fedoroff S Vernadakis A, eds.), Academic, Orlando, FL, pp. 293–310.

Dowling AJ Maggs A Scholes J (1991) Diversity amongst the microglia in growing and regenerating fish CNS: Immunohistochemical characterization using FL. 1, an anti-macrophage monoclonal antibody. Glia 4:345–364.

Easter SS Bratton B Scherer SS (1984) Growth-related order of the retinal fiber layer in goldfish. J Neurosci 4:2173–2190.

Easter SS Rusoff AC Kish PE (1981) The growth and organization of the optic nerve and tract in juvenile and adult goldfish. J Neurosci 1:793–811.

Ehinger B Zucker CL Bruun A Adolph A (1994) In vivo staining of oligodendroglia in the rabbit retina. Glia 10:40–48.

Fedoroff S Hao C (1991) Origin of microglia and their regulation by astroglia, in Plasticity and Regeneration of the Nervous System (Timiras PS Privat, A Giacobini E Lauder J Vernadakis A, eds.), Plenum, New York, pp. 135–161.

Fuchs C Druger RK Glasgow E Schechter N (1994) Differential expression of keratins in goldfish optic nerve during regeneration. J Comp Neurol 343:332–340.

Garrido M (1978) Estudio comparado del bulbo olfatorio de los vertebrados. Doctoral Thesis. Univ Sevilla.

Gayoso MJ Díaz-Flores L García JM Garrido M (1977) Factores que influyen en la morfología de los astrocitos. Morf Norm Patol Sec A 1:205–216.

Goetschy JF Ulrich G Aunis D Ciessielski–Treska J (1986) The organization and solubility properties of intermediate filaments and microtubules of cortical astrocytes in culture. J Neurocytol 15:375–387.

Hájos F Kálmán M (1989) Distribution of glial fibrillary acidic protein (GFAP)-immunoreactive astrocytes in the rat brain. II. Mesencephalon, rhomben-cephalon and spinal cord. Exp Brain Res 78:164–173.

Jeserich G (1981) A morphological and biochemical study of myelinogenesis in fish brain. Dev Neurosci 4:373–381.

Jeserich G Müller A Jacque C (1990) Developmental expression of myelin proteins by oligodendrocytes in the CNS of trout. Dev Brain Res 51:27–34.

Jeserich G Stratmann A (1992) In vitro differentiation of trout oligodendro-cytes: evidence for an A2B5-positive origin. Dev Brain Res 67:27–35.

Jeserich G Wachneldt TV (1986) Characterization of antibodies against major fish CNS myelin proteins: immunoblot analysis and immunohistochemi-cal localization of 36k and IP2 proteins in trout nerve tissue. J Neurosci Res 15:147–158.

Jeserich G Wachneldt TV (1987) Antigenic sites common to major fish myelin glycoproteins (IP) and to major tetrapods PNS myelin glycoprotein (Po) reside in the amino acid chains. Neurochem Res 12:821–825.

Kalnins VI Subrahmanyan L Opas M (1986) The cytoskeleton, in Astrocytes, vol. 3 (Fedoroff S Vernadakis A, eds.), Academic, Orlando, FL, pp. 27–60.

Kosaka T Hama K (1983) Synaptic organization in the teleost olfactory bulb. J Physiol, Paris 78:707–719.

Kruger L Maxwell DS (1966) The fine structure of the ependymal processes in the teleost optic tectum. Am J Anat 119:479–498.

Kruger L Maxwell DS (1967) Comparative fine structure of vertebrate neuro-glia. Teleosts and reptiles. J Comp Neurol 128:115–142.

Lara JM (1982) Estructura y ultraestructura del techo óptico de la carpa, *Cyprinus carpio*. Doctoral Thesis. Univ Sevilla.

Lara JM Aijón J (1983a) The optic tectum of the carp: Pyramidal neurons of the SFGS. J Hirnforsch 24:607–612.

Lara JM Aijón J (1983b) Myelinated nerve cell bodies in the carp olfactory bulb. Morf Norm Patol Sec A 7:677–688.

Lara JM Alonso JR Miguel JJ Aijón J (1987) Dense osmiophilic material in the surface of the olfactory bulb in the teleost *Cyprinus carpio* L. J Hirnforsch 28:233–235.

Lara JM Alonso JR Vecino E Coveñas R Aijón J (1989) Neuroglia in the optic tectum of teleosts. J Hirnforsch 30:465–472.

Laufer M Vanegas H (1974) The optic tectum of a perciform teleost. II. Fine structure. J Comp Neurol 154:69–96.

Lenhossék M von (1891) Zur Kenntnis der Neuroglia des menschlichen Rückenmarkes. Verh Anat Ges 5:193–221.

Levine RL (1989) Organization of astrocytes in the visual pathways of the gold-fish: An immunohistochemical study. J Comp Neurol 258:231–245.

Levine RL (1991) Gliosis during optic fiber regeneration in the goldfish: An immunohistochemical study. J Comp Neurol 312:549–560.

Levine RL (1993) Axon dependent glial changes during optic fiber regenera-tion in the goldfish. J Comp Neurol 333:543–553.

Lotan M Schwartz M (1992) Postinjury changes in platelet-derived growth fac-tor-like activity in fish and rat optic nerves. J Neurochem 58:1637–1642.

Maggs A Scholes J (1990) Reticular astrocytes in the fish optic nerve: Macroglia with epithelial characteristics form an axial repeated lacework pattern to which nodes of Ranvier are apposed. J Neurosci 10:1600–1614.

Markl J Frank WW (1988) Localization of cytokeratins in tissue of the rainbow trout: Fundamental differences in expression pattern between fish and higher vertebrates. Differentiation 39:97–122.

Miguel JJ Lara J Alonso JR Aijón J (1986) Structural organization of the optic tectum of Barbus meridionalis Risso. I. Inner strata (SPV SAC and SGC). J Hirnforsch 27:19–27.

Miller RH Ffrench–Constant C Raff MC (1989) The macroglial cells of the rat optic nerve. Annu Rev Neurosci 12:517–534.

Nona SN Duncan A Stafford CA Maggs A Jeserich G Cronly-Dillon JR (1992) Myelination of regenerated axons in goldfish optic nerve by Schwann cells. J Neurocytol 21:391–401.

Nona SN Shehab SAS Stafford CA Cronly-Dillon JR (1989) Glial fibrillary acidic protein (GFAP) from goldfish: Its localization in visual pathway. Glia 2:189–200.

Onteniente B Kimura H Maeda T (1983) Comparative study of the glial fibrillary acidic protein in vertebrates by PAP immunohistochemistry. J Comp Neurol 215:427–436.

Osborn M Debus E Weber K (1984) Monoclonal antibodies specific for vimentin. Eur J Cell Biol 34:137–143.

Penfield W (1924) Oligodendroglia and its relation to classical neuroglia. Brain 47:430–452.

Pinching AJ (1971) Myelinated dendritic segments in the monkey olfactory bulb. Brain Res 29:133–138.

Pinching AJ Powell TPS (1971) The neuron types of the glomerular layer of the olfactory bulb. J Cell Sci 9:305–345.

Privat A Rotaboul P (1986) Fibrous and protoplasmic astrocytes, in Astrocytes, vol. 1 (Fedoroff S Vernadakis A, eds.), Academic, Orlando, FL, pp. 105–129.

Quitschke W Schecher N (1984) 58,000 Dalton intermediate filament proteins of neuronal and nonneuronal origin in the goldfish visual pathway. J Neurochem 42:569–576.

Raff MC; Miller RH Noble M (1983) A glial progenitor cell that develops in vitro into an astrocyte or oligodendrocyte depending on culture medium. Nature 303:390–396.

Ramón y Cajal S (1890) Sur l'origene et les ramifications des fibers nerveuses de la moelle embryonnaire. Anat Anz 5:85–95.

Ramón y Cajal S (1911) Histologie du système nerveux de l'homme et des vertébrés. Maloine, Paris.

Ransch B Clapshaw PA Price J Noble M Seifert W (1982) Development of oligodendrocytes and Schwann cells studies with a monoclonal antibody against galactocerebroside. Proc Natl Acad Sci USA 79:2709–2713.

Rehn B Brepohl W Mendoza AS Apfelbach R (1986) Changes in granulle cells of the ferret olfactory bulb associated with imprinting on prey odours. Brain Res 373:114–125.

Reyners H Gianfelici de Reiners E Maisin JR (1982) The beta astrocyte: A newly recognized radiosensitive glial cell type in the cerebral cortex. J Neurocytol 11:967–983.

Rickmann M Amaral DG Cowan WM (1987) Organization of radial glial cells during the development of the rat dentate gyrus. J Comp Neurol 264:449–479.

Rio Hortega P (1921) Estudio sobre la neuroglía. La glia de escasas ramificaciones (Oligodendroglía). Bol Real Soc Españ Biol 9:69–120.

Roots BI (1986) Phylogenetic Development of astrocytes, in Astrocytes, vol. 1 (Fedoroff S Vernadakis A, eds.), Academic, Orlando, FL, pp. 1–34.

Rusoff AC Easter SS (1980) Order in the optic nerve of goldfish. Science 208:311–312.

Schmidt-Kastner R Ietasch K Weigel H Eysel UT (1993) Immunohistochemical staining for glial fibrillary acidic protein (GFAP) after deafferentation or ischemic infarction in rat visual system: features of reactive and damaged astrocytes. Int J Devel Neurosci 11:157–174.

Schmidt-Kastner R Szymas J (1990) Immunohistochemistry of glial fibrillary acidic protein (GFAP), vimentin and S-100 protein for study of astrocytes in hippocampus of rat. J Chem Neuroanat 3:179–192.

Sivron T Cohen A Hirschberg DL Jeserich G Schwartz M (1991) Soluble factor(s) produced in injured fish optic nerve regulate the postinjury number of oligodendrocytes: possible role of macrophages. Glia 4:591–601.

Sivron T Jeserich G Nona S Schwartz M (1992) Characteristics of fish glial cells in culture: possible implications as to their lineage. Glia 6:52–66.

Skoff RP Knapp PE Bartlett WP (1986) Astrocytic diversity in the optic nerve: A cytoarchitectural study, in Astrocytes, vol. 1 (Fedoroff S Vernadakis A, eds.), Academic, Orlando, FL, pp. 269–291.

Sommer I Schachner M (1981) Monoclonal antibodies (O1 to O4) to oligodendrocyte cell surfaces: An immunocytological study in the central nervous system. Dev Biol 83:311–327.

Sprinkle TJ (1989) 2',3'-Cyclic nucleotide 3'-phosphodiesterase an oligodendrocyte-Schwann cell and myelin-associated enzyme of the nervous system. Crit Rev Neurobiol 4:235–301.

Stafford CA Shehab SAS Nona SN Cronly-Dillon JR (1990) Expression of glial fibrillary acidic protein (GFAP) in goldfish optic nerve following injury. Glia 3:33–42.

Stevenson JA Yoon MG (1982) Morphology of radial glia, ependymal cells, and periventricular neurons in the optic tectum of goldfish *(Carassius auratus)*. J Comp Neurol 205:128–138.

Studnicka FK (1900) Untersuchungen über den Bau des Ependyms der nervösen Zentralorgane. Anat Hefte 15:301–431.

Tigges M Tigges J (1980) Distribution and morphology of myelinated perikarya and dendrites in the olfactory bulb of primates. J Neurocytol 9:825–834.

Trimmer PA Reier PJ Oh TH Eng LF (1982) An ultrastructural and immunocytochemical study of astrocytic differentiation in vitro. J Neuroimmunol 2:235–260.

Vanegas H Laufer M Amat J (1974) The optic tectum of a perciform Teleost. I. General configuration and cytoarchitecture. J Comp Neurol 154:43–60.

Vaughan DW (1984) The structure of neuroglial cells, in Cerebral Cortex, vol. 2 (Jones EG Peters A, eds.), Plenum, New York, pp. 285–329.

Vaugh JE Hinds PL Skoff PR (1970) Electron microscopic study of Wallerian degeneration in rat optic nerve. I. The multipotential glia. J Comp Neurol 140:175–206.

Vaugh JE Peters A (1968) A third neuroglial cell type. An electron microscopic study. J Comp Neurol 133:269–288.

Vauyiouklis DA Brophy PJ (1993) Microtubule-associated protein MAP1B expression precedes the morphological differentiation of oligodendrocytes. J Neurosci Res 35:257–267.

Velasco A (1992) Glia en el sistema nervioso central de teleósteos. Aportaciones morfológicas e inmunocitoquímicas. Minor Thesis. Univ Salamanca.

Virchow R (1846) Ueber das granulierte Ansehen der Wandungen der Gehiruventrike. Allg Z Psychiatr 3:224–250.

Waehneldt TV Matthieu JM Jeserich G (1986) Appearance of myelin proteins during vertebrate evolution. Neurochem Int 9:463–474.

Willey TJ (1973) The ultrastructure of the cat olfactory bulb. J Comp Neurol 152:211–232.

Wolburg H (1978) Growth and myelination of goldfish optic nerve fibers after retina regeneration and nerve crush. Z naturforsch 33:988–996.

Wolburg H (1981) Myelination and remyelination in the regenerating visual system of the goldfish. Exp Brain Res 43:199–206.

Wolburg H Bouzehouane U (1986) Comparison of the glial investment of normal and regeneration fiber bundles in the optic nerve and tectum of the goldfish and Crucian carp. Cell Tissue Res 244:187–192.

In Vitro Studies of Astrocyte Development in Higher Mammals

Gregory A. Elder

1. Introduction

Morphological studies of fixed brain sections have yielded many clues concerning astrocyte development in vivo in higher mammals. However, purely descriptive studies provide at most circumstantial evidence about lineage relationships. For example, it is still not known precisely when during development neuroectodermal cells become committed to neuronal or glial cell lineages. Recently recombinant retroviral tracers have been used to address this issue in the rat and chick embryo (Turner and Cepko, 1987; Galileo et al., 1990). Another approach more applicable especially to human studies is to address these issues in dissociated cell culture. As in other species, glial cells of various central nervous system (CNS) regions from humans and other higher mammals can be established in primary cultures and their development monitored in vitro. The utility of the in vitro approach was greatly advanced in the 1980s with the development of a variety of cell type specific antibodies (Raff et al., 1979), which were demonstrated to be useful for identifying not only fully differentiated cells but also early progenitors and cells in intermediate stages of development.

This chapter reviews information about astrocyte development in higher mammals from studies in dissociated cell culture. More specifically, attempts to apply markers used in delineating glial cell lineages in the rat CNS to higher mammals will be

From: *Neuron-Glia Interrelations During Phylogeny: I. Phylogeny and Ontogeny of Glial Cells*
A. Vernadakis and B. Roots, Eds. Humana Press Inc., Totowa, NJ

reviewed, with particular reference to the question of whether we can identify an equivalent of the rat O2-A cell lineage in higher mammals. By the term "higher mammals" the author will principally be referring to human studies, since less information exists for other species. A review of this topic seems warranted since assumptions are often made that cell lineage markers useful in the rat CNS will automatically mark equivalent cells in other mammals. Unfortunately, the number of studies directly addressing this issue remains small even in humans. However, for some markers it already appears that some rules that apply to rat glial development do not apply in other mammalian systems. Besides its significance to the basic biology of glial cell development in higher mammals, such studies may also be relevant to understanding astrocytic responses in human neurological disease. Additionally, the identification of markers for progenitor cells in the human CNS may have practical implications in isolating purified populations of progenitors for use in human transplantation.

2. The Rat O2-A Cell Lineage

The usefulness of the in vitro approach to studying glial cell lineages has probably best been demonstrated by Martin Raff and his collaborators in the identification of the rat O2-A cell lineage (for reviews, *see* Miller et al., 1989; Raff, 1989; Richardson et al., 1990). Originally Raff and his colleagues, while working with cultures of neonatal rat optic nerve, found that they could identify two populations of astrocytes based on differences in morphology, time of developmental appearance, and responses to mitogens (Raff et al., 1983a). Type 1 astrocytes appeared flatter and more fibroblast-like and only rarely labeled with a monoclonal antibody (A2B5) known to bind to polysialogangliosides (Eisenbarth et al., 1979; Kasai and Yu, 1983; Kundu et al., 1983) present on multiple cell types in the nervous system, including neurons and glia (Schnitzer and Schachner, 1982). Type 2 astrocytes were more neuron-like in appearance and nearly always labeled with A2B5 (Raff et al., 1983a). Developmentally, type 1 astrocytes appeared first, followed by oligodendrocytes and then type 2 astrocytes (Miller et al., 1985; Williams et al., 1985).

Unexpectedly, these cultures were also found to contain an A2B5+ bipotential progenitor cell (the O-2A cell) that could differentiate into a type 2 astrocyte in serum containing media or into an oligodendrocyte under serum-free conditions (Raff et al., 1983b). Although it had long been assumed that a common progenitor for astrocytes and oligodendrocytes existed, the suggestion that a separate astrocyte lineage might exist that was more related to oligodendrocytes than to other astrocytes represented a novel concept. Latter experiments showed that the processes of both oligodendrocytes and type 2 astrocytes structurally contribute to the nodes of Ranvier and lead to the concept that the O2-A cell is a progenitor of a cell lineage specialized for myelination (ffrench-Constant and Raff, 1986).

The O2-A cell is now known to be a rapidly proliferating progenitor labeled by several antibodies that recognize cell surface antigens (A2B5, GD3, LB1) (Curtis et al., 1988; Miller et al., 1989) and by expression of the intermediate filament protein vimentin (Raff et al., 1984). As O-2A cells mature they proliferate more slowly, and, if committed to become oligodendrocytes, express antigens recognized by the monoclonal antibody O4. O4 staining allows the identification of immature oligodendrocytes since it appears after precursors have begun to express A2B5 and GD_3, but before expression of Gal C or other myelin proteins (Dubois-Dalcq, 1987; Levi et al., 1987; Gard and Pfeiffer, 1990). Following Gal C expression, oligodendrocytes remain O4+ but A2B5 and GD_3 staining is rapidly lost both in cell culture (Levi et al., 1987) and in vivo (Curtis et al., 1988).

Several growth factors regulate O2-A cell differentiation. Platelet derived growth factor (PDGF) stimulates O-2A cell proliferation until an intrinsic timing mechanism causes progenitors to differentiate into oligodendrocytes (Lillien et al., 1988; Raff et al., 1988; Richardson et al., 1990). Basic fibroblast growth factor (bFGF) in combination with PDGF sustains O2-A proliferation without differentiation (Bogler et al. 1990), whereas insulin like growth factor-1 (IGF-1) induces oligodendrocyte differentiation (McMorris et al., 1986; McMorris and Dubois-Dalcq, 1988).

Type 1 astrocytes, the first glial cells to develop in rat optic nerve, arise during late embryonic life from a lineage distinct from the O2-A cell lineage. They, however, are thought to influence the development of oligodendrocytes and type 2 astrocytes by being an in vivo source of PDGF and beginning in the second postnatal wk, a protein similar to ciliary neurotrophic factor (CNTF) that causes remaining progenitors to begin differentiation into type 2 astrocytes (Lillien et al., 1988).

These observations on rat glial development initially based on experiments with rat optic nerve also appear to apply to differentiation in whole brain as well (Williams et al., 1985; Levi et al., 1987). In addition, an O2-A cell has been recognized in adult brain (Noble and Wolswijk, 1992; Wolswijk and Noble, 1989). The adult O-2A progenitor expresses both the A2B5 and O4 antigens but does not express vimentin. Although these cells can become oligodendrocytes or type 2 astrocytes in vitro, they differentiate more slowly in culture than neonatal O2-A cells.

3. Is There an Equivalent of the Rat O2-A Cell in Higher Mammals?

Many groups have shown that glial cell cultures can be established from humans and other higher mammals using methods often similar to those used to prepare rat glial cultures. For example, "shake off" methods can be used to prepare purified astrocyte cultures from human fetal brain cultures initially containing a mixed cell population (Kennedy and Fok-Seang, 1986; Major et al., 1990). In addition, methods exist for preparing highly purified cultures of human, ovine, or bovine oligodendrocytes in bulk cultures (Farooq et al., 1981; Elder et al., 1988; Kim, 1990). These cultures have been utilized by a number of groups to determine the susceptibility of neural cells to viral infections (see, for example Kennedy et al., 1983; Elder and Potts, 1987; Major and Vacante, 1989). In the case of human tissue, access is generally limited to fetal material of less than 24 wk gestational age or surgical specimens most often from adults. Access to second or third trimester human fetal material is rare.

Several early studies showed that the same cell type specific antibodies used to identify rat neural cells in culture were equally useful in identifying human astrocytes, oligodendrocytes, macrophages, and neurons in culture (Dickson et al., 1982; Kennedy et al., 1980). Since the general features of CNS development are similar in all mammals it would perhaps be expected to find functional equivalents of the rat O2-A cell lineage in higher mammals. However, only by performing analogous in vitro experiments with glial cultures from higher mammals can the utility of markers for rat glial progenitors be established in other species. To date a small series of studies has been published addressing this issue in human fetal optic nerve (Kennedy and Fok-Seang, 1986), human fetal brain (Elder and Major, 1988) or spinal cord (Aloisi et al., 1992b), adult human white matter (Armstrong et al., 1992), and fetal ovine brain (Elder et al., 1988).

3.1. Fetal Optic Nerve

Kennedy and Fok-Seang (1986) published the first study that compared rat and human glial development based on the O2-A cell model. They dissociated human optic nerve from 16–18-wk-old fetuses and then stained the cultured cells with A2B5 and other cell-type specific markers. As in rat optic nerve, they found that two populations of GFAP+ astrocytes could be distinguished using the A2B5 antibody. However, unlike rat, the morphological distinctions between human A2B5+ and A2B5- astrocytes were not as marked as in cultures of rat optic nerve. Human cultures contained both A2B5+ astrocytes with a fibroblast-like appearance as well A2B5- astrocytes with many processes.

In addition to the population of A2B5+ astrocytes many A2B5+ cells were GFAP- and GalC-. Frequently, these cells had a bipolar morphology resembling rat O2-A cells. More important, these cells responded to growth in serum containing (DMEM-FCS) or serum-free (DMEM-BS) media in a manner similar to rat O2-A cells. After 3 d in either culture condition, the appearance of A2B5+ astrocytes was favored in DMEM-FCS, whereas the numbers of oligodendrocytes was increased in DMEM-BS. In cultures in DMEM-FCS, A2B5+/GFAP+ cells increased nearly three fold from 9% on day 1 to 30% of total cells on d 3, whereas A2B5+/GFAP- cells decreased proportionately. During this same time, oligodendrocyte numbers changed little in DMEM-FCS. The number of GalC+ oligodendrocytes found in either serum-free or serum-containing conditions was small, yet even after 24 h, cultures in DMEM-BS contained nearly three times as many GalC+ cells as cultures grown in DMEM-FCS (0.7% in DMEM-FCS vs 2% in DMEM-BS). By d 3, a much larger increase in the number of GalC+ oligodendrocytes had occurred in cultures in DMEM-BS (8%) than was seen with DMEM-FCS (1.1%). Although A2B5+/GFAP+ cells also increased in DMEM-BS (5% on d 1 to 13% on d 3), the increase was less than in DMEM-FCS.

Complement-mediated lysis experiments showed that A2B5+ cells were required for the development of GalC+ oligodendrocytes and at least some A2B5+ astrocytes. No GalC+ cells developed in DMEM-BS following treatment with A2B5 antibody and complement, arguing that as in rat optic nerve, A2B5+ cells were needed for oligodendrocyte development. However, in contrast to the complete elimination of oligodendrocytes, A2B5+ astrocytes in DMEM-FCS fell by only about 50% compared to control cultures treated with complement alone. It was not clear whether this resulted from new A2B5+ cells being generated from an initially A2B5- population or incompleteness of the complement-mediated lysis of a subpopulation of cells.

Kennedy and Fok-Seang (1986) additionally prepared cultures of purified astrocytes from human fetal cortex by shaking off nonastrocytic cells after 10–12 days in vitro (DIV). As in rat cortical cultures prepared by similar methods, the purified human astrocytes consisted of greater than 95% GFAP + cells that were uniformly A2B5-. The purified human astrocytes like rat astrocytes expressed the neural cell adhesion molecule (NCAM) and also supported the unfaciculated growth of rat CNS neurons.

Taken together, these results suggested mostly similarities between cultured rat and human optic nerve cells. Most important, they pointed to the existence of an A2B5+ cell the equivalent of the rat O2-A progenitor in human optic nerve.

3.2. Fetal Brain and Spinal Cord

Subsequent studies, although not contradicting any of the data gathered from human fetal optic nerve, have indicated that some aspects of the rat O2-A model, particularly the use of the Mab A2B5 may not be applicable to glial development in other brain regions in higher mammals. While examining cultured human fetal brain at gestational ages younger (7–10 wk) than those studied by Kennedy and Fok-Seang (1986), we too found A2B5+ and A2B5- astrocytes (Elder and Major, 1988). However, we were surprised to find many A2B5+ astrocytes (presumptive human type 2) in cultures of early human fetal brain that contained few A2B5- astrocytes (presumptive human type 1 astrocytes) and no oligodendrocytes. This observation suggested that A2B5+ astrocytes, which appear last of these three cell types in rat brain, were appearing earlier than A2B5-astrocytes and oligodendrocytes in human brain.

We next compared the cellular composition of short-term cultures and fresh cell suspensions from human fetal brain established from 8-10 and 16–18-wk-old fetal material (Fig. 1). After 48 h in vitro, astrocytes constituted 8–14% of the cells in culture. In 8–10-wk fetal brain, over 90% of the astrocytes were A2B5+. In cultures from 16–18-wk brain, most astrocytes were still A2B5+, although there was a relative increase in the proportion of A2B5- astrocytes (33%) to A2B5+ (67%) astrocytes. As in human optic nerve, A2B5+ and A2B5- astrocytes in fetal brain cultures could not be distinguished reliably based on morphology (Fig. 2). By the first passage of cultures established from 16-wk fetal brain, over 90% of the astrocytes were A2B5-, whereas cultures containing nearly equal numbers of A2B5- and A2B5+ astrocytes developed from cultures of 8–10-wk brain. Such a rapid transformation suggested that either a selective loss of A2B5+ astrocytes must occur with time in culture or, more likely, that A2B5- astrocytes proliferate more rapidly than the A2B5+ component. This latter behavior may represent one similarity of human A2B5- and A2B5+ astrocytes to rat type 1 and type 2 astrocytes, since rat type 1 astrocytes also proliferate more rapidly in culture than type 2 astrocytes. These results are also consistent with Kennedy and Fok-Seang's (1986) finding that astrocytes in passaged cultures from 24-wk brain are nearly all A2B5-. Cultures of fetal brain from these ages contained no GalC+ oligodendrocytes and only rare neurofilament+ neurons.

To confirm that the ratios of human astrocytes subsets observed in culture corresponded to those in vivo, we determined the proportions of human A2B5+ and A2B5- astrocytes in fresh cell suspensions of 7–8- and 15–16-wk gestational age human fetal brain (Fig. 1). Nearly 80% of astrocytes in the 7–8-wk brain cell suspensions were A2B5+, whereas nearly equal numbers of A2B5- and A2B5+ cells were found in suspensions from 15–16-wk fetal brain. The relatively larger numbers of A2B5- astrocytes in fresh cell suspensions compared with the short-term cultures may reflect a partial loss of the A2B5 epitope during the dissociation procedure with its subsequent regeneration in vitro. Alternatively, some conversion of A2B5+/GFAP- cells into A2B5+/GFAP+ cells may have occurred in culture. In either case, the data obtained by staining fresh cell suspensions confirm that most astrocytes in early human fetal brain are A2B5+, whereas A2B5- astrocytes increase

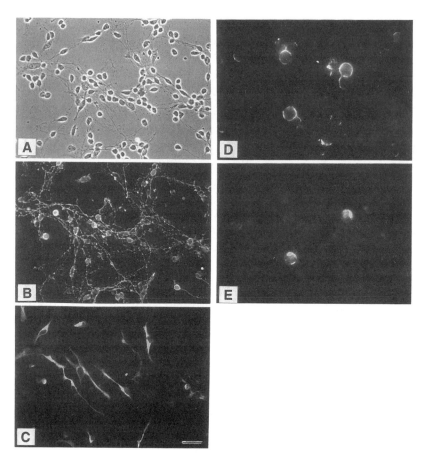

Fig. 1. Human fetal brain cultures after 10 DIV, phase contrast **(A)** or after immunolabeling with A2B5 **(B)** or GFAP **(C)**. Note the fine membranous processes of the many A2B5+/GFAP- cells. Most of the GFAP+ cells are also labeled with A2B5. In **D** and **E** fresh cell suspensions from 8-wk fetal brain were labeled with A2B5 (D) or GFAP (E). Both GFAP+ cells are also A2B5+. For experimental details, *see* Elder and Major (1988). Bar = 20 μm (A-C), 10 μm (D-E).

latter. A2B5- astrocytes are also the predominate type cultured from adult human brain (Newcombe et al., 1988; Yong et al., 1990).

Within the A2B5+ cell population were GFAP- cells with rounded cell bodies and often bipolar processes similar to the rat O2-A like cells found in human optic nerve cultures. However, these cells lacked vimentin-intermediate filaments, and no induc-

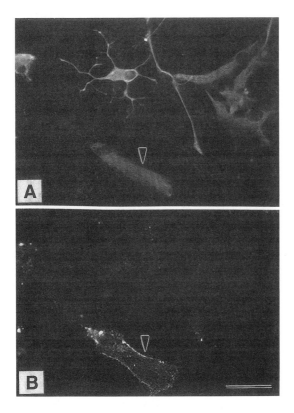

Fig. 2. Astrocytes from 14 wk human fetal brain in culture immuno-labeled for GFAP **(A)** or A2B5 **(B)**. One flat fibroblast like astrocyte is A2B5+ (indicated by an arrowhead), whereas a stellate shaped cell and several flatter GFAP+ cells are A2B5-. Bar = 20 μm.

tion of Gal C+ oligodendrocytes was achieved when cultures of either human fetal spinal cord or cerebrum were maintained in a serum-free medium identical to that used for inducing differentiation of rat O-2A cells into oligodendrocytes. Thus, although cells morphologically and in some respects immunocytochemically similar to the rat O2-A cell occur in human fetal brain these cells could not be shown to be bipotential.

A general question remains as to the nature of the many undifferentiated cells found in these cultures that do not label with antibodies to GFAP, GalC, or neurofilament. These cells, which are relatively homogenous morphologically, generally have small round cell bodies, often with bipolar processes. They are the predominant cell type in cultures of human fetal brain from gestational ages 8–24 wk and may be divided into an A2B5+ and A2B5- population. Lee et al. (1992), in a further analysis of human brain cultures from fetuses of ages 16–24 wk, also found that these cells failed to stain with most antibodies, including glial and neuronal markers. The cells most consistently stained with A2B5 but did not label with antibody to the ganglioside GD_3, a marker for primitive subventricular neuroepithelial cells *(see later)*. Lee et al. (1992), concluded based on immunocytochemical and EM analysis that these cells were most likely immature postmitotic neurons. However, in cultures from younger fetuses (7–8 wk-old) many of these cells are clearly not postmitotic (unpublished observations) and may well represent germinal neuroepithelial cells.

In a separate study, we also examined the cellular composition of cultures from the developing ovine CNS (Elder et al., 1988). For these studies, cultures from 50-d fetal (full gestation in sheep is 145 d) to adult animals were characterized. Many similarities were found with the studies of cultured human fetal brain. As in rats and humans, both A2B5+ and A2B5- astrocytes could be identified in ovine cultures and, as in humans, the ovine A2B5+ and A2B5- could not be reliably differentiated by their morphology, which was more influenced by whether the cells were maintained in serum-free or serum-containing medium than by their A2B5 labeling status (Fig. 3). In addition, ovine A2B5+ astrocytes were present in cultures form early fetal brain before the development of identifiable oligodendrocytes. At the earliest fetal age examined (25 d) nearly all astrocytes were A2B5+, although a small A2B5- population could be identified.

A2B5+ cells with a small round cell body often with bipolar processes were also identifiable. These cells could be isolated in large numbers from the cerebellum of animals up to 140 d gestational age, although only rarely were such cells seen in cultures derived from the cerebral hemispheres after 75 d gestational age. When cultures from 100- to 115-d ovine cerebellum were maintained in 5% serum, most cells under went a morphological transition to a flatter fibroblastic morphology. As the morphological transition occurred, an increasing fraction of A2B5+ cells began expressing GFAP. When similar cultures were placed in serum-free media, although the A2B5+ cells assumed a more oligodendroglial-like appearance they were not induced to express Gal C.

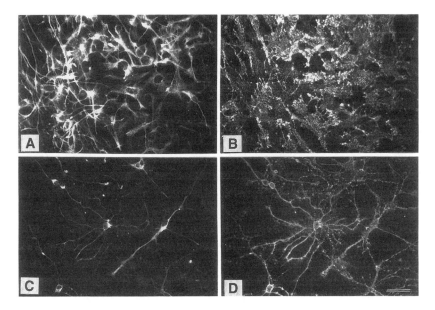

Fig. 3. Ovine cerebellar cultures from 110-d fetal animal at 10 DIV, cultured in 5% serum **(A, B)** or serum-free media **(C, D)**. Cultures are dual labeled for GFAP (A, C) or A2B5 (B, D). Cells in serum-free media contain more rounded cell bodies and a more process-bearing morphology than cells in serum. For experimental details, *see* Elder et al. (1988).

Thus, a population of ovine A2B5+/GFAP- cells appears capable of differentiating into A2B5+ astrocytes, but could not be shown to be bipotential under these culture conditions.

More recent studies (Aloisi et al., 1992b) have documented the developmental appearance of glial cell types in human spinal cord cultures from 6–9-wk-old fetuses. In addition to A2B5, Aloisi et al. (1992b) used antibodies to the ganglioside GD_3, SSEA-1, and the presumptive early oligodendrocyte marker O4. Antibody to GD_3, a ganglioside principally expressed in the developing nervous system (Seyfried and Yu, 1985) has been shown by Goldman and his colleagues to label neuronal and glial precursors in the subventricular zone of the cerebrum and the external germinal layer of the cerebellum. (Goldman et al., 1984; Goldman et al., 1986). In vitro, GD3 antibody labels rat cells that can differentiate both into astrocytes and oligodendrocytes, although the cells become GD3- following differentiation (Goldman et al., 1986).

In human spinal cord, no GFAP+ cells were found in short term cultures established from 6–7-wk-old fetal material. GFAP+ cells first appeared in cultures from 8–9-wk fetal spinal cords, but represented less than 3.0% of cultured cells. All astrocytes in 8-wk cultures were vimentin+ and more than 90% were A2B5+, GD_3+, and SSEA-1+. Thus, as in whole brain, almost all GFAP+ cells at this age were A2B5+ and additionally nearly all expressed the GD_3 marker.

Among the GFAP- cells a large population (about 40%) labeled with both A2B5 and GD_3, whereas 20% of the total cells were SSEA-1+. Most GFAP- cells had a small round cell bodies and later developed longer, thinner processes that could become fasciculated. Only a few stained with antineurofilament antibodies.

O4, a presumptive early oligodendrocyte marker, labeled cells present in cultures established from ages as early as the seventh wk. At 8 wk, 2–3% of cells were O4+ and a few O4+ cells expressed Gal C. During the first wk in culture, many O4+ cells had bi- or tri-polar morphologies and incorporated ^3H thymidine findings consistent with these cells being proliferating precursor cells. However, no actual differentiation of O4+/Gal C- cells into Gal C+ cells was demonstrated.

3.3. Progenitor Cells in Adult Brain

The existence of O2-A progenitor cells in adult rat brain has also stimulated a search for equivalent cells in the brains of higher mammals. Armstrong et al. (1992) established cultures from adult human white matter taken from patients undergoing partial temporal lobe resections for intractable epilepsy. They found that after 1–2 wk in vitro these cultures contained a population of process-bearing cells that were partially similar in their immunocytochemical profile to O2-A progenitors in adult rat optic nerve. The human cells were O4+ but did not express GalC, GFAP, or vimentin. Unlike adult rat O-2A progenitors, the human O4+ cells did not stain with A2B5 or LB1.

Besides putative progenitor cells (O4+/Gal C-/GFAP-) these cultures also contained O4+/Gal C+ oligodendrocytes, cells termed type 2 astrocytes based on their O4 positivity (O4+/ GalC-/GFAP+) and cells with a mixed oligodendrocyte/astrocyte phenotype (O4+/GalC+/GFAP+). These latter two categories of cells were only rarely observed.

A cell population analysis suggested that the O4+/GalC-/ GFAP- cells could differentiate into oligodendrocytes. O4+ cells were most abundant at 1 wk in vitro but then decreased in number by 2 wk, and had nearly disappeared by 3 wk in culture. In contrast, oligodendrocyte numbers generally increased between 1 and 3 wk in culture. Armstrong et al. (1992) demonstrated that elimination of O4+ cells from the cultures by complement-mediated lysis inhibited oligodendrocyte development supporting the notion that O4+ is a marker for preoligodendrocytes in adult human white matter. Evidence for the in vivo existence of a human adult progenitor cells came from the finding of cells with the preoligodendrocyte (O4+/GalC-/GFAP-) antigenic profile in tissue prints of adult human white matter, although relative to oligodendrocytes the O4+ progenitors were relatively rare.

Unlike rat O2-A cells human preoligodendrocytes did not divide in response PDGF, IGF-1, or bFGF. However the relative ratios of oligodendrocytes to preoligodendroctyes was increased by IGF-1, whereas the opposite effect was produced by bFGF treatment. Thus in adult progenitors IGF-1 likely induces oligodendrocyte differentiation and bFGF seems to inhibit this process.

Two additional studies have suggested the existence of astrocyte progenitors in adult human or bovine brain. Norton and Farooq (1989) prepared cultures of adult bovine white matter by a method that produces highly purified populations of oligodendrocytes. After 1 DIV they found that the cultures contained only GalC+/GD$_3$- oligodendrocytes (90–95% of the cells), and small round cells that were GD3+/Gal C- (4–10%). However, when maintained in serum-free media, these cultures evolved to contain a mixture of oligodendrocytes and astrocytes, even though no GFAP+ cells had been present initially. They followed the expression of GD$_3$ as a possible marker of immature astrocyte progenitors. Based on the evolution of the staining pattern, they concluded that round GD$_3$+/GFAP- cells gave rise to small process-bearing GD$_3$+/vimentin+/GFAP- cells that then rapidly proliferated and became GFAP+. Although some GFAP+ astrocytes remained GD$_3$+, most appeared to lose the GD$_3$ antigen before differentiating. No evidence was found to indicate that any cells were differentiating into oligodendrocytes under these conditions, all of which were GD$_3$-. This latter finding contrasts with the results of Kim et al. (1986), who found that adult human oligodendrocytes in culture were GD_3+. Whether these contrasting results reflects species differences in GD_3 expression or differences in staining conditions is not known.

Finally Perzel'ova and Mares (1993) studied explant cultures from adult human brain biopsies obtained from both cortical and white matter regions. The primary cultures were composed of rapidly proliferating GFAP-/vimentin+ cells. However, with prolonged passage, a spontaneous growth deceleration occurred accompanied by the *de novo* appearance of GFAP+ cells. No other markers were used to assess the nature of the cells involved in this evolution.

3.4. Use of Other Markers

Rat type I astrocytes label with antibodies to the glycoprotein Ran-2 (Miller et al., 1989), whereas rat type 2 astrocytes are also labeled with HNK-1 (L2), NG-2 (Miller et al., 1989), and by antibodies to the extracellular matrix protein janusin (Bartsch et al., 1993). To date there has not been any systematic attempt to use these antibodies to study cultured glial cells in higher mammals. Subpopulations of human astrocytes isolated from adult brain also stain with antibodies to gangliosides G_{M1} or G_{M4} (Kim, 1990) and the intercellular adhesion molecule-1 (ICAM-1) (Satoh et al., 1991). MHC class II expression has been observed on a restricted population of astrocytes either under basal or induced conditions (Grenier et al., 1989). In addition, morphological heterogeneity of human adult astrocytes in culture has also been correlated with HLA-DR expression (Yong et al., 1990). Whether any of these antigens will serve as useful lineage markers remains unexplored.

4. Conclusions Regarding In Vitro Studies

At this point one must admit that relatively little is known about glial cell lineages in higher mammals. The studies published to date are mostly observational in nature. Only a few have used techniques, such as complement mediated lysis, for addressing differentiation in vitro, and no studies have analyzed isolated progenitors in single-cell culture systems, as rat O2-A cells have been analyzed. Without doubt in the case of human studies this situation is largely owing to the limited availability of human fetal material from a range of developmental ages. Fetuses from a broader range of developmental ages potentially are available from other large animals, such as sheep or nonhuman primates. However, studies in these species are constrained by economic considerations and likely technical considerations as well.

To date the most compelling evidence for an A2B5+ bipotential O2-A cell in humans comes from the data generated with human optic nerve (Kennedy and Fok-Seang, 1986). A2B5 does seem to be a marker of a subset of astrocyte progenitors in sheep (Elder et al., 1988) and likely also human brain (Elder and Major, 1988). Likewise, O4 appears to be a marker of oligodendrocyte precursors (Armstrong et al., 1992) and GD$_3$ of astrocyte precursors (Norton and Farooq, 1989) in adult brain. However, none of these latter progenitors appear to be bipotential in vitro.

Although the terms type 1 and type 2 astrocyte continue to be used occasionally to describe A2B5- and A2B5+ astrocytes in higher mammals, except perhaps in optic nerve the utility of this designation is in doubt. The phenomenon of A2B5+ and A2B5- astrocytes seems to be widespread in higher mammals, such subtypes being found not only in humans but also in nonhuman primates (unpublished observations) and sheep. However the presence of human A2B5+ astrocytes early in brain development before the appearance of oligodendrocytes seems well established and suggests that the monoclonal antibody A2B5 cannot be used to delineate a cell lineage equivalent to the rat O2-A in whole brain. Until these issues are better clarified in higher mammals, the designations type 1 and type 2 astrocyte probably should be avoided since the terms imply both lineage and functional relationships and not simply immunocytochemical profiles.

5. Correlation of In Vitro and In Vivo Data

One factor complicating comparisons between rat and human in vitro glial studies is the fact that the human fetal brain used in most studies represents a much earlier developmental stage than neonatal rat brain or optic nerve. The most significant developmental changes occurring between the 7th and 16th wk of gestation in human cerebrum is the proliferation and migration of cells away from the primitive ventricular zone (Sidman and Rakic, 1982). Neurogenesis occurs mostly later, whereas myelin is not detectable in spinal cord until around 20 wk and only later in cerebrum (Richardson, 1982). Another factor to consider is the vastly different time span of development in the two species. In humans, neuronal and glial differentiation occurs over long time periods, especially in areas such as neocortex where apparently undifferentiated cells persist in the germinal matrix until late in gestation. This might offer one explanation for why it has been so difficult to identify bipotential glial progenitors in vitro. If at any point in time only a small fraction of precursor cells are in the process of differentiating relatively few cells may be in a state allowing continued differentiation in culture or the time course may be prolonged. It might also mean that different developmental rules are needed to achieve differentiation in higher mammals.

The radial glia, which function as guides for neuronal migration, are the first glial cells to become recognizable in primate brain (Rakic, 1984; *see also* the chapter by Voight and de Lima). In human spinal cord, radial glia can be recognized morphologically at 6 wk of age (Choi, 1981). Initially, radial glia express vimentin (Lukas, et al. 1989). By 8–9 wk they start to express GFAP (Choi, 1981; Lukas et al., 1989), being the first cells to exhibit GFAP reactivity in fetal spinal cord. In the human hippocampus, the first glial cells to appear are also vimentin-positive radial glial (Janus et al., 1991), with the transition from vimentin to GFAP expression occurring at 8 wk gestational age. Thus, the A2B5+ astrocytes which are the predominate astrocyte type in cultures established from fetuses of 8–10 wk gestational age, may correspond to human radial glial cells. Interestingly radial glia in embryonic rat spinal cord are also A2B5+ (Hirano and Goldman, 1988). Janus et al. (1991) have also identified a population of small S-100 positive cells that were GFAP- and vimentin- and morphologically distinct from radial glial cells. These cells are found as early as 9.5 wk in hippocampus and, since S-100 is generally regarded as an astrocytic marker, these cells may represent an additional astrocyte lineage present in early fetal brain development.

Radial glial undergo a second proliferative period and subsequently are believed to be transformed into astrocytes (Schmechel and Rakic, 1979a,b) and possibly oligodendrocytes (Choi et al., 1983; Hirano and Goldman, 1988). Choi et al (1983) have studied this process in human spinal cord and based on morphological and immunocytochemical observations, suggested that radial glia become both astrocytes and oligodendrocytes. They have observed in human cord of 13- and 16-wk cells with mixed astrocyte/oligodendrocyte and immature oligodendrocyte-like morphological features that have GFAP immunoreactivity (Choi and Kim, 1984, 1985).

The transformation of radial glia into astrocytes in human hippocampus occurs after 14 wk gestational age (Janus et al., 1991). At this same time, a transition toward a population of A2B5- astrocytes is seen in both short term cultures and fresh cell suspensions of human fetal cerebrum. It is not known whether this represents a transformation of A2B5+ astrocytes into A2B5- cells or the emergence of a separate A2B5- astrocyte lineage.

6. Functional Correlations of Astrocyte Subtypes

At present few correlations can be drawn between the heterogeneity of astrocytes demonstrated by antibody markers and the functional subgroups of astrocytes that likely exist in vivo. The rat type 2 astrocyte appears to serve important functions in myelination (ffrench-Constant and Raff, 1986), whereas the growth factors produced by type 1 astrocyte likely regulate O2-A cell differentiation (Lillien et al., 1988; Richardson et al., 1990). How type 1 and type 2 astrocytes in vitro relate to protoplasmic and fibrous astrocytes in vivo is unclear. The possible correlation between A2B5+ astrocytes early in development and radial glia has been mentioned already.

The heterogeneity revealed by the immunological markers, however, likely represents only the beginning of identifying the real diversity of astrocytes in brain based on function. Beside a structurally supportive role in the CNS, there is now considerable evidence that astrocytes perform an array of functions that may vary regionally. Astrocytes are suspected of playing roles in processes as diverse as axonal guidance (Silver et al., 1982; Hankin and Silver, 1988), induction of the blood-brain barrier (Janzer and Raff, 1987; Joo, 1987), and even regulation of cerebral blood flow (Paulson and Newman, 1987). The expression of immune markers on human fetal astrocytes suggests roles in immunological functions in the CNS (Ennas et al., 1992). Likewise, human astrocytes either constituitively express or can be induced to secrete a variety of growth factors, cytokines, and immune system modulators (Aloisi et al., 1992a; Johnson et al., 1993; Lee et al., 1993; Sharif et al., 1993), including tumor necrosis factor-α, transforming growth factor-β, interleukins 1, 6, and 8, and colony stimulating factor-1, the latter a growth factor that selectively promotes the proliferation, survival, and differentiation of cells of the mononuclear phagocytic series. It has recently been observed that in vitro human fetal astrocytes produce a factor midkine, which promotes the survival of cultured mesencephalic neurons (Kikuchi et al., 1993; Satoh et al., 1993). These cytokines and growth factors may function during development or in response to injury. The secretion of a variety of metalloproteinases by astrocytes may also be important in facilitating cellular migration during development or in inflammatory responses in the CNS (Apodaca et al., 1990).

Astrocytes play important roles in regulating the ionic composition of the extracellular fluid of the brain (Landis, 1994). Several neurotransmitters can be taken up by astrocytes, including glutamate and GABA. Neurotransmitter uptake may be particularly important in glutamatergic transmission since astrocytes contain the enzyme glutamine synthetase (Martinez-Hernandez et al., 1977). In brain, expression of this enzyme is restricted to astrocytes, suggesting that astrocytes play a role in returning glutamine to neurons by taking up synaptically released glutamate and converting to glutamine.

Interestingly astrocytes can also be depolarized by some neurotransmitters through ligand gated receptors (Bowman and Kimelberg, 1984; Kettenmann et al., 1984) including certain subtypes of glutamate receptors (von-Blankenfeld and Kettenmann, 1991). Although it is not clear that astrocytes regulate information processing in neuronal circuits, their responses to neurotransmitters are likely important in integrating neuronal and glial functions. However, some direct signaling between astrocytes and neurons in cultures has been suggested recently (Nedergaard, 1994).

Astrocytes may protect neurons in other ways as well. Human astrocytes contain the enzyme quinolinic acid phophorribosyltransferase the degradative enzyme of the endogenous excitotoxin quinolinic acid (Du et al., 1990). The presence of glutathione S-transferase (Carder et al., 1990), a complex group of multifunctional proteins that detoxify a range of toxic substances including drugs and carcinogens, suggests that astrocytes function in other detoxification processes as well.

Astrocytes respond to almost all types of injury or disease that can affect the CNS, including trauma, infections, neurotoxic damage, inflammatory demyelination, and genetic and degenerative diseases (Duchen, 1992). However, comparisons of gliosis in a variety of animal models has shown that gliotic reactions can vary widely in their time course and degree of hyperplastic responses, suggesting that different biological mechanisms may be operative depending on the nature of the insult (Norton et al., 1992). Besides reactive gliosis astrocytes may contribute even more fundamentally to the characteristic pathology or even the pathophysiology of some disorders. Astrocytes produce both the A/β peptide (Busciglio et al., 1993) as well as the serine protease inhibitor α 1 antichymotrypsin (Koo et al., 1991), two components of the amyloid deposits found in senile plaques and blood vessels in Alzheimer's disease. Interestingly, astrocytes are also the major source in brain of apolipoprotein E (Boyles et al., 1985), a component of senile plaques and neurofibrillary tangles that in certain of its isozyme forms recently has been discovered to be a major risk factor for late onset sporadic Alzheimer's disease (Corder et al., 1993).

Given these diverse functions, astrocytes may well represent far more lineages than are currently recognized. Hopefully, future research can relate these functions to separable astrocyte lineages, and despite the difficulties in working with fetal tissues from higher mammals, these relationships can be understood in higher mammals as well. Clearly, the discovery of additional markers that can aid in delineating glial lineages in higher mammals is needed.

References

Aloisi F Care A Borsellino G Gallo P Rosa S Bassani A Cabibbo A Testa U Levi G Peschle C (1992a) Production of hemolymphopoietic cytokines (IL-6, IL-8, colony-stimulating factors) by normal human astrocytes in response to IL-1beta and tumor necrosis factor alpha. J Immunol 149:2358–2366.

Aloisi F Giampaolo A Russo G Peschle C Levi G (1992b) Developmental appearance antigenic profile and proliferation of glial cells of the human embryonic spinal cord: an immunocytochemical study using dissociated cultured cells. Glia 5:171–181.

Apodaca G Rutka JT Bouhana K Berens ME Giblin JR Rosenblum ML McKerrow JH Banda MJ (1990) Expression of metalloproteinases and metalloproteinase inhibitors by fetal astrocytes and glioma cells. Cancer Res 50:2322–2329.

Armstrong RC Dorn HH Kufta CV Friedman E Dubois-Dalcq ME. (1992) Preoligodendrocytes from adult human CNS. J Neurosci 12:1538–1547.

Bartsch U Pesheva P Raff M Schachner M (1993) Expression of janusin (J1-160/180) in the retina and optic nerve of the developing and adult mouse brain. Glia 9:57–69.

Bogler O Wren D Barnett SC Land H Nobel M (1990) Cooperation between two growth factors promotes extended self-renewal and inhibits differentiation of oligodendrocyte-type-2 astrocyte (O2-A) progenitor cells. Proc Natl Acad Sci USA 87:6368–6372.

Bowman CL Kimelberg HK (1984) Excitatory amino acids directly depolarize rat brain astrocytes in primary culture. Nature 311:656–659.

Boyles JK Pitas RE Wilson E Mahley RW Taylor JM (1985) Apolipoprotein E associated with astrocytic glia of the central nervous system and with nonmyelinating glia of the peripheral nervous system. J Clin Invest 76:1501–1513.

Busciglio J Gabuzda DH Matsudaira P Yankner BA (1993) Generation of beta-amyloid in the secretory pathway in neuronal and nonneuronal cells. Proc Natl Acad Sci USA 90:2092–2096.

Carder PJ Hume R Fryer AA Strange RC Lauder J Bell JE (1990) Glutathione S-transferase in human brain. Neuropathol Appl Neurobiol 16:293–303.

Choi BH (1981) Radial glia of developing human fetal spinal cord: Golgi, immunohistochemical and electron microscopic study. Dev Brain Res 1:249–267.

Choi BH Kim RC (1984) Expression of glial fibrillary acidic protein in immature oligodendroglia. Science 223:407–409.

Choi BH Kim RC (1985) Expression of glial fibrillary acidic protein in immature oligodendroglia and its implications. J Neuroimmunol 8:215–235.

Choi BH Kim RC Lapham LW (1983) Do radial glia give rise to both astroglia and oligodendroglial cells? Dev Brain Res 8:119–130.

Corder EH Saunders AM Strittmatter WJ Schmechel DE Gaskell PC Small GW Roses AD Haines JL Pericak-Vance MA (1993) Gene dosage of apolipoprotein E type 4 allele and the risk of Alzheimer's disease in late onset families. Science 261:921–923.

Curtis R Cohen J Fok-Seang J Hanley MR Gregson MA Reynolds R Wilkin GP (1988) Development of macroglial cells in rat cerebellum I: Use of antibodies to follow early in vivo development and migration of oligodendrocytes. J Neurocytol 17:43–54.

Dickson JG Flanigan TP Walsh FS (1982) Cell-surface antigens of human fetal brain and dorsal root ganglion cells in tissue culture. Adv Neurol 36:435–451.

Du F Okuno E Whetsell WO Kohler C Schwarcz R (1990) Distribution of quinolinc acid phosphoribosyltransferase in the human hippocampal formation and parahippocampal gyrus. J Comp Neurol 295:71–82.

Dubois-Dalcq M (1987) Characterization of a slowly proliferating cell along the oligodendrocyte differentiation pathway. EMBO J 6:2587–2595.

Duchen LW (1992) General pathology of neurons and neuroglia, in Greenfield's Neuropathology (Adams JH Duchen LW, eds.), Oxford University Press, New York, pp. 34–46.

Eisenbarth GS Walsh FS Nirenberg M (1979) Monoclonal antibody to a plasma membrane antigen of neurons. Proc Natl Acad Sci USA 76:4913–4917.

Elder GA Potts BJ (1987) Multiple neural cell types are infected in vitro by border disease virus. J Neuropathol Exp Neuro 46:653–667.

Elder GA Major EO (1988) Early appearance of type II astrocytes in developing human fetal brain. Devel Brain Res 42:146–150.

Elder GA Potts BJ Sawyer M (1988) Characterization of glial subpopulations in cultures of the ovine central nervous system. Glia 1:317–327.

Ennas MG Cocchia D Silvetti E Sogos V Riva A Torelli S Gremo F (1992) Immunocompetent cell markers in human fetal astrocytes and neurons in culture. J Neurosci Res 32:424–436.

Farooq M Cammer W Snyder DS Raine CS Norton WT (1981) Properties of bovine oligodendroglia isolated by a new procedure using physiologic conditions. J Neurochem 36:431–440.

ffrench-Constant C Raff MC (1986) The oligodendrocyte-type 2 astrocyte cell lineage is specialized for myelination. Nature 323:335–338.

Galileo DS Gray GE. Owens GC Majors J Sanes JR (1990) Neurons and glia arise from a common progenitor in chicken optic tectum: Demonstration with two retroviruses and cell type specific antibodies. Proc Natl Acad Sci USA 87:458–462.

Gard AL Pfeiffer SE (1990) Two proliferative stages of the oligodendrocyte lineage (A2B5+O4- and O4+GalC-) under different mitogenic controls. Neuron 5:615–625.

Goldman JE Hirano M Yu RK Seyfried TN (1984) GD3 ganglioside is a glycolipid characteristic of immature neurochemical cells. J Neuroimmunol 7:179–192.

Goldman JE Geier SS Hirano M (1986) Differentiation of astrocytes and oligodendrocytes from germinal matrix cells in primary culture. J Neurosci 6:52–60.

Grenier Y Ruijs TCG Robitaille Y Olivier A Antel JP (1989) Immunohistochemical studies of adult human glial cells. J Neuroimmunol 21:103–115.

Hankin MH Silver J (1988) Development of intersecting CNS fiber tracts: The corpus callosum and its perforating fiber pathway. J Comp Neurol 272:177–190.

Hirano M Goldman JE (1988) Gliogenesis in rat spinal cord: Evidence for origin of astrocytes and oligodendrocytes from radial glial precursors. J Neurosci Res 21:155–167.

Janus MS Nowakowski RS Mollgard (1991) Glial cell differentiation in neuron-free and neuron rich regions II: Early appearance of S-100 protein positive astrocytes in human fetal hippocampus. Anat Embryol Berl 184:559–569.

Janzer RC Raff MC (1987) Astrocytes induce blood brain barrier properties in endothelial cells. Nature 325:253–257.

Johnson MD Jennings MT Gold LI Moses HL (1993) Tranforming growth factor-beta in neural embryogenesis and neoplasia. Hum Pathol 24: 457–462.

Joo F (1987). Current aspects of development of the blood-brain barrier. Int J Dev Neurosci 5:369–372.

Kasai N Yu RK (1983) The monclonal antibody A2B5 is specific to ganglioside GQ1c. Brain Res 277:155–158.

Kennedy PGE Lisak RP Raff MC (1980) Cell type-specific markers for human glial and neuronal cells in culture. Lab Invest 43:342–351.

Kennedy PGE Clements GB Brown SM. (1983) Differential susceptibility of human neural cell types in culture to infection with herpes simplex virus. Brain 106:101–119.

Kennedy PGE Fok-Seang J (1986) Studies on the development antigenic phenotype and function of human glial cells in tissue culture. Brain 109: 1261–1277.

Kettenmann H Backus KH Schachner M (1984) Aspartate, glutamate, and gamma-aminobutyric acid depolarize cultured astrocytes. Neurosci Lett 52:25–29.

Kikuchi S Muramatsu H Muramatsu T Kim SU (1993) Midkine, a novel neurotrophic factor, promotes survival of mesencephalic neurons in culture. Neurosci Lett 160:9–12.

Kim SU Moretto G Lee V Yu RK (1986) Neuroimmunology of gangliosides in human neurons and glial cells in culture. J Neurosci Res 15:303–321.

Kim SU (1990) Neurobiology of human oligodendrocytes in culture. J Neurosci Res 27:712–728.

Koo EH Abraham CR Potter H Cork LC Price DL (1991) Developmental expression of alpha 1-antichymotrypsin in brain may be related to astroglisis. Neurobiol Aging 12:495–501.

Kundu SK Pleatman MA Redwine WA Boyd AE Marcus DM (1983) Binding of monoclonal antibody A2B5 to gangliosides. Biochem Biophys Res Commun 116:836–842.

Landis DMD (1994) The early reactions of non-neuronal cells to brain injury, in Annual Review of Neuroscience, vol. 17 (Cowan WM, ed.), Annual Reviews Inc., Palo Alto, CA, pp. 133–151.

Lee SC Liu W Brosnan CF Dickson DW (1992) Characterization of primary human fetal dissociated central nervous system cultures with an emphasis on microglia. Lab Invest 67:465–476.

Lee SC Liu W Roth P Dickson DW Berman JW Brosnan CF (1993) Macrophage colony stimulating factor in human fetal astrocytes and microglia. Differential regulation by cytokines and lipopolysaccharide and modulation of class II MHC on microglia. J Immunol 150:594–604.

Levi G Aloisi F Wilkin GP (1987) Differentiation of cerebellar glial precursors into oligodendrocytes in primary culture: developmental profile of surface antigens and mitotic activity. J Neurosci Res 18:407–417.

Lillien LE Sendtner M Rohrer H Hughes SM Raff MC (1988) Type-2 astrocyte development in rat brain cultures is initiated by a CNTF-like protein produced by type-1 astrocytes. Neuron 1:485–494.

Lukas Z Draber P Bucek J Draberova E Viklicky V Staskova Z (1989) Expression of vimentin and glial fibrillary acidic protein in human developing spinal cord. Histochem J:21 693–702.

Major EO Vacante DA (1989) Human fetal astrocytes in culture support the growth of the neurotropic human polyomavirus, JCV. J Neuropathol Exp Neurol 48:425–436.

Major EO Amemiya K Elder G Houff SA (1990) Glial cells of the human developing brain and B cells of the immune system share a common DNA binding factor for recognition of the regulatory sequences of the human polyomavirus, JCV. J Neurosci Res 27:461–471.

Martinez-Hernandez A Bell KP Norenberg MD (1977). Glutamine synthetase: glial localization in brain. Science 195:1356–1358.

McMorris FA Dubois-Dalcq M (1988) Insulin-like growth factor 1 promotes cell proliferation and oligodendroglial commitment in rat glial progenitor cells developing in vitro. J Neurosci Res 21:199–209.

McMorris FA Smith TM DeSalvo S Furlanetto RW (1986) Insulin-like growth factor I/somatomedin C: a potent inducer of oligodendrocyte development. Proc Natl Acad Sci USA 83:822–826.

Miller RH David S Patel R Abney ER Raff MC (1985) A quantitative immunohistochemical study of macroglial cell development in the rat optic nerve: In vivo evidence for two distinct astrocyte lineages. Dev Biol 111:35–41.

Miller RH ffrench-Constant C Raff MC. (1989) The macroglial cells of the rat optic nerve, in Annual Review of Neuroscience, vol. 12 (Cowan WM, ed.), Annual Reviews Inc., Palo Alto CA, pp. 517–534.

Nedergaard M (1994) Direct signaling from astrocytes to neurons in cultures of mammalian brain cells. Science 263:1768–1771.

Newcombe J Meeson A Cuzner ML (1988) Immunocytochemical characterization of primary glial cell cultures from normal adult human brain. Neuropathol Appl Neurobiol 14:453–465.

Noble M Wolswijk G (1992) Development and regeneration in the O2-A lineage: studies in vitro and in vivo. J Neuroimmunol 40:287–294.

Norton WT Farooq M (1989). Astrocytes cultured from mature brain derive from glial precursor cells. J Neurosci 9:769–775.

Norton WT Aquino DA Hozumi I Chiu F-C Brosnan CF (1992) Quantitative aspects of reactive gliosis: a review. Neurochem Res 17:877–885.

Paulson OB Newman EA (1987) Does the release of potassium from astrocyte end-feet regulate cerebral blood flow? Science 237:896–898.

Perzel'ova A Mares V (1993) Appearance of GFAP-positive cells in adult brain cultures spontaneously decelerated in growth. Glia 7:237–244.

Raff MC Fields KL Hakomori SI Mirsky SI Pruss RM Winter J (1979) Cell-type specific markers for distinguishing and studying neurons and the major classes of glial cells in culture. Brain Res 174:283–308.

Raff MC Abney ER Cohen J Lindsay R Noble M (1983a) Two types of astrocytes in culture of developing rat white matter: Differences in morphology, surface gangliosides and growth characteristics. J Neurosci 3:1289–1300.

Raff MC Miller RH Noble M (1983b) A glial progenitor cell that develops in vitro into an astrocyte or an oligodendrocyte depending on culture medium. Nature 303:390–396.

Raff MC Williams BP Miller RH (1984) The in vitro differentiation of a bipotential glial progenitor cell. EMBO J 3:1857–1864.

Raff MC Lillien LE Richardson WD Burne JF Noble M (1988) Platelet-derived growth factor from astrocytes drives the clock that times oligodendrocyte development in culture. Nature 333:562–565.

Raff MC (1989) Glial cell diversification in the rat optic nerve. Science 243:1450–1455.

Rakic P (1984) Emergence of neuronal and glial cell lineages in primate brain, in Cellular and Molecular Biology of Neuronal Development (Black IB, ed.), Plenum, New York, pp. 29–50.

Richardson EP (1982) Myelination in the human central nervous system, in Histology and Histopathology of the Nervous System (Haymaker W Adams RD, eds.), Thomas, Springfield, IL, pp. 146–173.

Richardson WD Raff M Noble M (1990) The oligodendrocyte-type 2 astrocyte lineage. Semin Neurosci 2:445–454.

Satoh J Kastrukoff LF Kim SU (1991) Cytokine-induced expression of intercellular adhesion molecule-1 (ICAM-1) in cultured human oligodendrocytes and astrocytes. J Neuropathol Exp Neurol 50:215–226.

Satoh J-I Muramatsu H Moretto G Muramatsu T Chang HJ Kim ST Cho JM Kim SU (1993) Midkine that promotes survival of fetal human neurons is produced by fetal human astrocytes in culture. Dev Brain Res 75:201–205.

Schmechel DE Rakic P (1979a) Arrested proliferation of radial glial cells during midgestation in rheuses monkey. Nature 277:303–305.

Schmechel DE Rakic P (1979b) A Golgi study of radial glial cells in developing monkey telencephalon: Morphogenesis and transformation into astrocytes. Anat Embryol 156:115–152.

Schnitzer J Schachner M (1982) Cell type specificity of neural cell surface antigen recognized by monoclonal antibody A2B5. Cell Tissue Res 224:625–636.

Seyfried TN Yu RK (1985). Ganglioside GD$_3$: Structure, cellular distribution, and possible function. Mol Cell Biochem 68:3–10.

Sharif SF Hariri RJ Chang VA Barie PS Wang RS Ghajar JB (1993) Human astrocyte production of tumour necrosis factor-alpha, interleukin-1, and interleukin-6 following exposure to lipopolysaccharide endotoxin. Neurol Res 15:109–112.

Sidman RL Rakic P. (1982) Development of the human central nervous system, in Histology and Histopathology of the Nervous System (Haymaker W Adams RD, eds.), Thomas, Springfield, IL, pp. 3–145.

Silver J Lorenz SE Wahlsten D Coughlin J (1982) Axonal guidance during development of the great cerebral commissures: Descriptive and experimental studies, in vivo, on the role of preformed glial pathways. J Comp Neurol 210:10–29.

Turner DL Cepko C (1987) A common progenitor for neuron and glia persists in the rat retina late in development. Nature 328:131–136.

von-Blankenfeld G Kettenmann H (1991). Glutamate and GABA receptors in vertebrate glial cells. Mol Neurobiol 5:31–43.

Williams BP Abney ER Raff MC (1985) Macroglial cell development in embryonic rat brain: studies using monclonal antibodies, fluorescence activated cell sorting, and cell culture. Dev Biol 112:126–134.

Wolswijk G Noble M (1989) Identification of an adult-specific glial progenitor cell. Development 105:387–400.

Yong VW Yong FP Olivier A Robitaille Y Antel JP (1990) Morphologic heterogeneity of human adult astrocytes in culture: correlation with HLA-DR expression. J Neurosci Res 27:678–688.

Schwann Cells in Phylogeny

Helen J. S. Stewart
and Kristjan R. Jessen

1. Introduction

Schwann cells are the nonneuronal cells that ensheath peripheral axons. Brought to prominence by Theodor Schwann in 1839 these cells have, until recently, been considered to be relatively passive components of the nerve, apart from their vital role in providing myelin sheaths around the larger axons. However, the Schwann cell has now come of age: recent studies have pointed to a role of Schwann cells as providers of growth factors and cell adhesion molecules that are permissive for regrowth of axons following nerve damage; in certain circumstances Schwann cells have the capacity to interact with the cells of the immune system, they take part in regulating the axonal microenvironment and provide extracellular matrix molecules for the mechanical strengthening of peripheral nerves. There is also rapid progress in analyzing the involvement of Schwann cells in peripheral nerve disorders, such as Charcot-Marie-Tooth disease, using the methods of molecular biology.

In this chapter we will document the evolution of Schwann cells and their invertebrate counterparts, and we will consider how Schwann cell–axon relationships and functions have changed during phylogeny.

2. Evolution of Peripheral Glia

To consider the evolution of peripheral glia it is first necessary to outline briefly the evolution of species (*see also* Roots, 1993). All the main phyla identifiable today originally evolved from

From: *Neuron-Glia Interrelations During Phylogeny: I. Phylogeny and Ontogeny of Glial Cells*
A. Vernadakis and B. Roots, Eds. Humana Press Inc., Totowa, NJ

primitive flagellate stock approx 600 million years ago during the Cambrian era (Fig 1). The most primitive phyla, the protozoa, and the porifera, lack nervous systems and thus lack glia. The coelenterates (sea anemones, hydra), however, have a primitive network of isolated neurons and naked processes but they lack peripheral glial cells, although there may be a few glia in the ganglia of jellyfish (Bullock and Horridge, 1965; Roots, 1978). Glial cells are also few in number in the marine phylum echinodermata (sea urchins, starfish) (Roots, 1986; Pentreath, 1987; Cobb, 1991) and the platyhelminths (flatworms) (Roots, 1986). Peripheral glia are, however, clearly apparent in the nerves of more complex phyla such as annelids, marine, fresh water, and earthworms, arthropods (insects, crabs, prawns, spiders and scorpions, and so on), and molluscs (snails, shellfish, squids, and so on). As the organizational complexity of the phyla increases, so too do the number of glial cells relative to neurons, and the association of the glia with the neurons also becomes more intricate *(see later)*. Glia are present in the peripheral nerves of all higher invertebrates and in vertebrates.

3. Types of Peripheral Glia

Most studies in invertebrates describe only one type of peripheral glial cell. In general, all nonneuronal cells associated with nerves are described as glia and few distinctions have been made between them. Thus the cells of the perineurium are referred to as glia, whereas in vertebrates these are recognized as a distinct cell type that, unlike other peripheral glia, develop from fibroblasts (Bunge et al., 1989). Where distinctions in invertebrate peripheral glia have been documented, three subtypes are seen: perineurial glia type I and II and subperineurial glia (Maddrell and Treherne, 1967; Blanco 1988). The subperineurial glia correspond to vertebrate Schwann cells in that they ensheath peripheral axons (Blanco, 1988).

In vertebrates, four distinct types of peripheral glia are seen. These are the enteric glia (the enteric ganglia intrinsic to the gut), teloglia (neuromuscular junction), satellite cells (sensory, sympathetic, and parasympathetic ganglia), and Schwann cells. Schwann cells can be described as cells that ensheath peripheral axons and they exist in vivo in two subtypes, myelin-forming cells and

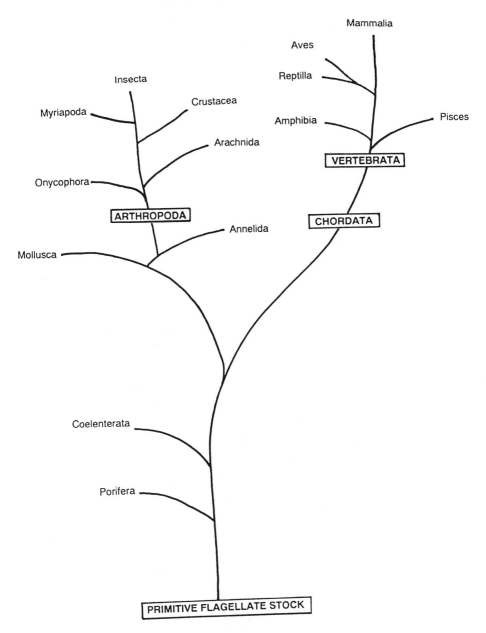

Fig. 1. Simplified phylogenetic tree showing the relationships between the main phyla mentioned in this chapter.

nonmyelin-forming cells. These Schwann cells differ in their relationship with axons and in their functional properties (*see below*, and Bray et al., 1981).

In this chapter, we will use the term Schwann cell to describe vertebrate Schwann cells and the subperineurial glia of invertebrate axons. Where distinctions in invertebrate peripheral glia types have not been made, we will refer to the cells as peripheral glial cells.

4. Origin of Schwann Cells

In mammals, using cell tracing techniques, it has been clearly demonstrated that Schwann cells are one of several cell types that originate from the neural crest (Le Douarin and Smith, 1988; Weston, 1991). Moreover, recent studies carried out in rodents show that the development of neural crest cells into Schwann cells proceeds via an intermediate cell type known as the Schwann cell precursor (Jessen et al., 1994). Although it is likely to exist, the presence of this cell type in nerves from vertebrates other than rat and mouse has not yet been directly demonstrated (*see later*).

The embryological origin of peripheral glia in lower phyla is less clear. In annelids and arthropoda these cells may be derived from the ectoderm (Roots, 1978) an observation supported by recent studies in the fruit fly, Drosophila melanogaster, that show that glial cells of the fly wing are clonally related to epithelial cells, derivatives of the ectoderm (Giangrande, 1994).

5. Evolution of the Glial Cell Sheath

As a general rule, axons of the peripheral nervous system are ensheathed by glial cells, although the complexity of the sheath varies with the increasing complexity of the animal as a whole (Fig. 2). Naked axons are found primarily in coelenterates but are sometimes found in certain places in higher animals including mammals (Bullock and Horridge, 1965). A common relationship between Schwann cells and axons is a simple ensheathing (Fig. 2B). This arrangement, where a single glial cell ensheaths many axons is frequently seen in annelida, molluscs, and arthropods (Pentreath, 1987, 1989; Blanco, 1988). Interestingly, a similar arrangement also is seen in embryonic nerves of higher vertebrates, such as rats (Peters and Muir, 1959; Jessen et al., 1994)

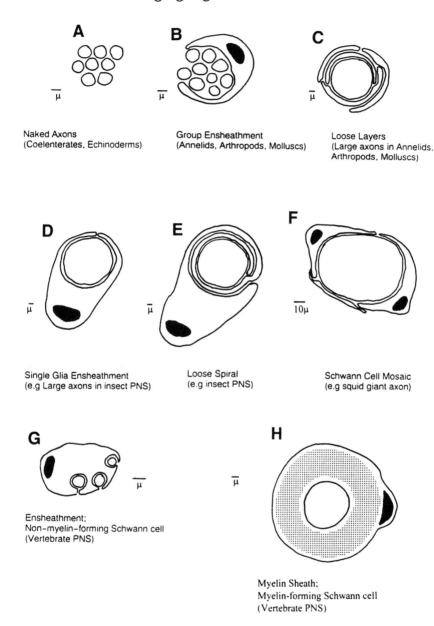

A

B

C

Naked Axons
(Coelenterates, Echinoderms)

Group Ensheathment
(Annelids, Arthropods, Molluscs)

Loose Layers
(Large axons in Annelids,
Arthropods, Molluscs)

D

E

F

Single Glia Ensheathment
(e.g Large axons in insect PNS)

Loose Spiral
(e.g insect PNS)

Schwann Cell Mosaic
(e.g squid giant axon)

G

H

Ensheathment;
Non-myelin-forming Schwann cell
(Vertebrate PNS)

Myelin Sheath;
Myelin-forming Schwann cell
(Vertebrate PNS)

Fig. 2. Sheath arrangements in different phyla. Adapted with permission from Pentreath et al., 1987.

(Fig. 3). In this case, each Schwann cell precursor takes part, with other precursors, in ensheathing hundreds of very small diameter axons. As the nerve grows and axonal diameter increases, the Schwann cell precursors mature into Schwann cells that gradually invade the axon bundles, eventually ensheathing a small group of axons individually, or wrapping a single large diameter axon to form the myelin sheath *(see later)* (Jessen and Mirsky, 1991; Jessen et al., 1994). Perhaps it is useful to regard the glia seen in peripheral nerves of these lower phyla as the counterparts of the Schwann cell precursor seen in higher vertebrate embryonic nerves today.

More intricate sheath arrangements have also been reported in several invertebrate phyla, e.g., arthropods (Fig. 2B,C). In the higher invertebrate phyla, loose sheaths consisting of many concentric lamellae, that resemble myelin when observed with the light microscope, are seen. However, studies have shown that these sheaths differ from vertebrate myelin in several ways, for example, the glial cell nucleus is in on the inside of the sheath and the laminae do not connect to form a spiral. In general, as in vertebrates, large diameter axons are surrounded by a greater number of glial wrappings than small diameter axons (Heuser and Doggenweiler, 1966). In contrast, the giant fibers of the squid (molluscs) are covered by a very thin mosaic of Schwann cells ($<1 \mu$ thick) that are only a few cell processes deep (Fig. 2F) (Bullock and Horridge, 1965).

In the evolutionarily more advanced phylum, chordata, in the vertebrates, two distinct sheath arrangements are seen in mature peripheral nerves. Single Schwann cells may wrap concentrically around large diameter ($>1 \mu$) axons to form the myelin sheath *(see later)* and are thus named myelin-forming Schwann cells, or a single Schwann cell may ensheath several small diameter axons, each axon generally lying in a separate surface trough; these Schwann cells can be referred to as nonmyelin-forming Schwann cells (Fig. 2G,H). In both of these sheath arrangements, axons are completely separated from each other.

6. Evolution of the Myelin Sheath

Although glial structures showing some similarities to myelin sheaths have been observed in the giant fibers of crustaceans *(see earlier)* (Heuser and Doggenweiler, 1966), true myelin sheaths are not apparent until the vertebrates. The myelin sheath is formed

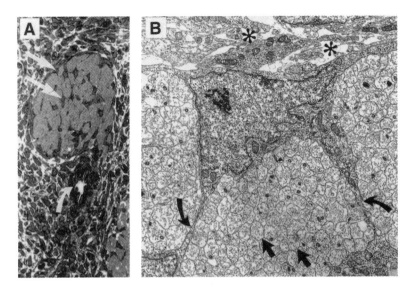

Fig. 3. The relationship of Schwann cell precursors to axons in early peripheral nerve development. The micrographs are of transverse sections of nerves in the hind limb of a 14-d-old rat embryo. **(A)** A semithin section showing one axon bundle surrounded by loosely arranged mesenchymal cells; a developing blood vessel is present below the nerve (curved arrow). Many process-bearing precursors are present within the nerve and at its edge (e.g., straight arrows). **(B)** One of the cells in the latter location is viewed by electron microscopy. On the top of the micrograph, large extracellular spaces are visible in the mesenchymal tissue (e.g., asterisks). The nerve itself is very compact: Numerous small axons of relatively uniform diameter (e.g., straight arrows) are separated by thin cytoplasmic sheets (e.g., curved arrows) that arise from the centrally located precursor cell and meet up with other precursors just visible at the bottom left and right corners. Reproduced with permission from Jessen et al., 1994.

by the concentric wrapping of a Schwann cell around a large diameter axon. The Schwann cell membranes are compacted, squeezing out cytoplasm, leaving restricted areas of cytoplasm mainly around the Schwann cell nucleus and in the part of the cell adjacent to the axon.

Compact myelin is found in all vertebrate classes except agnathes, i.e., lampreys and hagfish, the most primitive vertebrate class. Lamprey *(Petromyzon marinus)* nerves are surrounded by Schwann cells but no myelin is present (Peters, 1960; Bullock et al., 1984). The myelin sheath probably first appears among living vertebrates in the chondrichthyes (sharks and rays) that diverged from mammals about 400 million years ago (Romer, 1970).

Some of the proteins of peripheral myelin are also present throughout the vertebrates (except agnathes). The myelin proteins Po and myelin basic protein (MBP) have been reported in the PNS of shark (chrondrichthyes, cartilaginous fishes) (Saavedra et al., 1989), trout, toadfish (osteichthyes, bony fishes) (Jeserich and Waehneldt, 1986; Hammer et al., 1993), axolotl, bullfrog, African clawed toad (amphibia) (Takei and Uyemura, 1993), and mammals (Lemke, 1988). Shark Po and MBP protein are 46 and 44% homologous, respectively, to their mammalian counterparts (Saavedra et al., 1989).

7. Functions of Peripheral Glia: Phylogenetic Comparisons

It is apparent that glial cells are not merely passive components of the peripheral nerve but have numerous and varied functions that are essential for nerve development and maintenance. Some of these functions and how they have altered during evolution are described later.

7.1. Mechanical Support

A role for peripheral glia in providing mechanical support for nerves has been suggested in the lower invertebrate phyla. As species from these phyla often lack any skeletal support, the glial cells may act to prevent nerve damage that may occur owing to distortion during movement. Several structural observations support a mechanical function for invertebrate glia: Large numbers of intercellular junctions and adhesions are seen between glia, between glia and neurons, and between glia and the basement membrane (Lane, 1981; Blanco, 1988; Blanco and Lane, 1990). Furthermore, glial cells of molluscs and annelids have large numbers of filaments, whereas those of arthropods (which possess an exoskeleton) are relatively filament free, although they contain microtubules (Radojcic and Pentreath, 1979; Lane, 1981).

In vertebrates, where peripheral nerves contain large numbers of collagen fibers, a mechanical role for peripheral glia is less clear. It is of interest to note that nonmyelin-forming Schwann cells contain cytoplasmic filaments that are biochemically distinct from those in myelin-forming cells and that dye injection experiments show that only nonmyelin-forming Schwann cells are interconnected by gap junctions (Jessen et al., 1990; Konishi, 1990).

7.2. Impulse Conduction

Electrical impulses are conducted along axons to and from the periphery. In invertebrates, increases in the velocity of impulse conduction are generally achieved by increasing axon diameter. This results in the generation of very large axons, such as the giant axon of the squid, which can be up to 1 mm in diameter. In vertebrates, however, increased conduction velocities are mainly attained by the specialization of Schwann cells. Schwann cells wrap concentrically around large diameter axons to form the myelin sheath that acts as an insulating membrane. Along the length of an axon each myelinating Schwann cell is separated from the adjacent Schwann cell by a specialized area of electrically excitable axon surface known as a node of Ranvier. Action potentials leap from one node to the next by a process known as saltatory conduction, making impulse conduction down a myelinated fiber faster than down a continuously active unmyelinated one.

7.3. Ion Homeostasis

Glial cells probably play an important part in regulating the ionic environment of neurons. This may be particularly important in the case of K^+. Accumulation of K^+ in the extracellular space can lead to depolarization block of neuronal action potentials and affect synaptic transmission. Extracellular potassium levels can be controlled in a passive manner by spatial buffering (where glial cell membranes are selectively permeable to K^+) and/or by active pumping of K^+ into and Na^+ out of the cell using the Na-K-ATPase.

Passive K^+ clearance may be especially important in lower vertebrates and in insects (Kuffler, 1967; Gardner-Medwin et al., 1981; *see* Brown and Abbott, 1993 for discussion). Although similar mechanisms may exist in the vertebrate CNS (Gardner-Medwin, 1983), it is less clear how such a system would operate in the Schwann cells of peripheral nerve trunks of higher vertebrates, although Schwann cells from these species, including humans, possess voltage gated K^+ channels (Chiu, 1987; Bevan, 1990). It is more likely that in these species K^+ is to a larger extent cleared by active uptake into Schwann cells (Bevan, 1990) and, indeed, neurons. In contrast, there is little evidence for active K^+ uptake in invertebrate glia (Pentreath, 1987).

Thus glial cells from different phyla can regulate the ionic environment of neurons. The means by which they do so, however, is more sophisticated in more complex organisms.

7.4. Macromolecule Transfer Between Schwann Cells and Axons

It is widely believed, that glial cells interact metabolically with neurons in invertebrates, although evidence for a similar process occurring in vertebrate peripheral glial cells is limited. One of the best documented phenomena in this regard is the transfer of proteins from Schwann cells into the axoplasm in the squid giant axon. In this preparation it has been estimated that 20–40% of the proteins synthesized from radiolabeled precursors in Schwann cells is transferred (Lasek et al., 1974). It is interesting that this process may be specific, since, in some experiments, only certain glial proteins, including the 70 kDa heatshock protein and actin, were found to be translocated (Tytell et al., 1986). There is evidence for Schwann cell to axon transport of protein in other invertebrates including crayfish and aplysia. Furthermore, this exchange may not be unidirectional, since neuronally injected substances, such as horseradish peroxidase and lucifer yellow, have been observed to end up in the surrounding glia (Viancour et al., 1981; Goldstein et al., 1982). How transport of macromolecules from Schwann cells to axons occurs is unclear. The three following possibilities, or variants thereof, are frequently suggested. First, thin projections of Schwann cells seen in some preparations to extend into the axon interior are pinched off and internalized by the axon. Second, the Schwann cell may externalize whole vesicles that are subsequently taken up by the axon. Third, exocytosis by the Schwann cell might be tightly coupled to endocytosis in the adjacent part of the axonal membrane (for discussion, see Pentreath, 1987; Bittner, 1991; Buchheit and Tytell, 1992). In vertebrates, there is some evidence to suggest that glia to axon transfer occurs in the CNS. For example, astrocytes at the node of Ranvier may supply voltage-sensitive sodium channels to the axolemma (Black and Waxman, 1990). Schwann cells also synthesize voltage-sensitive sodium channels and it has been hypothesized that they are transferred to the axon (Gray and Ritchie, 1985). However, as the sodium channels of Schwann cells have different kinetic properties from those in axons, this hypothesis would require a change in channel properties once in the axolemma (Barres et al., 1990). The evidence for transfer of molecules between vertebrate Schwann cells and axons is, however, poor at present. It is possible that this feature of Schwann cells has become redundant during evolution.

7.5. Phagocytosis

Phagocytosis, although not a primary function of glia, may be important in removing redundant neuronal processes during development and in clearing up cellular debris and myelin following nerve damage. Although there is evidence for a phagocytic function of glia in invertebrates, most of the evidence is from studies on the CNS (Pentreath, 1987). In higher vertebrates, Schwann cells are probably involved in phagocytosing and degrading myelin debris after nerve injury, although they will not initiate phagocytosis in the absence of macrophages (Perry and Brown, 1992). Furthermore, after nerve injury, Schwann cells acquire aspects of a macrophage-related phenotype as they express the galactose specific lectin MAC-2, a molecule that is characteristically expressed by mature inflammatory macrophages (Rotshenker et al., 1992).

In addition, Schwann cells may assist phagocytosis in injured nerves by secreting chemoattractants for macrophages. For example, studies have shown that Schwann cells synthesize the potent macrophage chemoattractant Il1 in response to bacterial antigens (Bergsteinsdóttir et al., 1991).

7.6. Transmitter Uptake/Release/Inactivation

There is evidence for glial involvement in transmitter uptake and release in both vertebrates and invertebrates although, in higher vertebrates, most of the information comes from studies on CNS glia rather than Schwann cells. Some of the most intriguing observations in this area come from the squid. Schwann cells in squid synthesize and store acetylcholine that is released in response to an activity dependent axonal signal, the nature of which is still unclear. Released acetylcholine then acts via a feedback mechanism on Schwann cell nicotinic receptors causing an increase in K^+ conductance and an extended hyperpolarization of the cells (Villegas, 1981, 1984). What the functional significance of this process may be is currently unclear.

Uptake of other transmitters has been reported in invertebrate peripheral glia. Glutamate is actively taken up by glia in crustacea peripheral nerves and uptake of GABA occurs in glial cells of the lobster neuromuscular junction (Orkand and Kravitz, 1971; Evans, 1974). In vertebrates, it is clear that acetylcholine is synthesized and released by Schwann cells of the frog, provided the cells have been denervated by nerve transection (Katz and Miledi, 1959; Bevan et al., 1973). Although uptake of GABA has been noted in rat satellite cells and enteric glial cells (Schon and Kelly, 1974; Jessen et al., 1979) there appear to be few detailed studies on transmitter uptake in Schwann cells of higher vertebrates.

8. Comparison of Schwann Cells in the Vertebrates

The majority of studies on vertebrate Schwann cells have been carried out using mammals, particularly rats and mice. However, we will review the similarities and differences between vertebrate Schwann cells where possible.

8.1. Schwann Cell Development

Schwann cell development has been studied in only a few vertebrate species. In humans, the ulnar nerve of 6-wk-old (menstrual age) fetuses contains glial cells that communally ensheath axons, but do not express the Schwann cell marker protein S100. Furthermore, the ulnar nerve is at this stage not surrounded by discernible perineurium and blood vessels do not penetrate the nerve. S100 immunoreactivity starts to appear in the cells of the nerve around the seventh week (Gamble, 1966; Shearman and Franks, 1987). In these respects the cells and developmental stage of the 6 wk fetal human nerve resemble the cells recently identified as Schwann cell precursors in peripheral nerves of 14–15-d-old rat (Fig. 3) embryos (Jessen et al., 1994).

Communal envelopment of axons still prevails at 10 wk, whereas the first myelinated axons are seen in the ulnar nerve in foetuses of about 12 wk of age (Gamble, 1966) and about 2 wk later in the sciatic nerve (Cravioto, 1965). Thus the first traces of myelination are seen in human nerves at the end of the first quarter of embryonic life, whereas in the rat and mouse myelination does not occur until around birth (Webster, 1971; Brown and Asbury, 1981).

The development of human Schwann cells at 12 wk of life is approx equivalent to that of rat Schwann cells of 19.5–20.5 days of life (Peters and Muir, 1959; Gamble and Breathnach, 1965). It must be noted, however, that different nerves were used in these studies and that a rostral caudal gradient of nerve development exists.

As in man, myelination in the chick peripheral nerve begins well before birth and is first seen in 14-d-old embryos (Geren, 1954; Uyemura et al., 1979). It is intriguing that expression of the main myelin protein P_0 is seen much earlier and is even present in the neural crest (Bhattacharyya et al., 1991). It remains to be seen whether this is the case in mammals.

8.2. Schwann Cell Phenotypic Markers

In the rat sciatic nerve, numerous studies have mapped the Schwann cell phenotype during development and differentiation and in vitro (Jessen and Mirsky, 1991). Although limited studies have been carried out in other vertebrates, the molecular phenotype of Schwann cells has been well conserved, in general, in different vertebrate orders. A comparison of the molecular phenotype of Schwann cells from different vertebrate classes is shown in Table 1.

The low affinity nerve growth factor receptor (p75 NGFR) is expressed by human and rat Schwann cells during early development and in culture (Assouline and Pantazis, 1989). At later developmental times, however, downregulation of p75 NGFR occurs on both Schwann cell types but it is still detectable on nonmyelin-forming cells (Yan and Johnson, 1987; Scarpini et al., 1988; Jessen et al., 1990). The calcium binding protein S100 is also expressed throughout Schwann cell development in human and rat (*see above*).

Myelin protein expression by Schwann cells in vertebrates is more clearly documented than that of other phenotypic markers. Comparative studies on myelin proteins are dealt with in Chapter 11 in this volume.

Table 1
A Comparison of the Molecular Phenotype of Schwann Cells
from Various Vertebrate Species[a]

Antigen	Rat	Human	Frog	Chicken
S100	+	+	+	+
NCAM	+	+	nd	+
p75NGFR	+	+	nd	+
GalC	+	+	nd	nd
04	+	nd[b]	nd	+
GAP-43	+	nd	+	nd
Vimentin	+	+	nd	nd
GFAP	+	+	nd	nd
Po	+	+	nd	+

[a]This table illustrates the expression of molecular phenotype of Schwann cells from different vertebrate classes. For a more detailed analyzis of the molecular phenotype of rat Schwann cells, see Jessen and Mirsky, 1990. Note: As limited data are available, no attempt has been made to indicate at what developmental ages these antigens are expressed or how they are distributed between myelin-forming and nonmyelin-forming Schwann cells. References. Jessen and Mirsky, 1991; Bhattacharyya et al., 1991; Golding and Tonge, 1993; Morrisey et al., 1991; Miyazaki et al., 1994; Zimmerman and Suter, 1983; Scarpini et al., 1986, 1988; Kim et al., 1989; Bianchini et al., 1992; Shearman and Franks, 1987; Curtis et al., 1992; Jauberteau et al., 1992).
[b]nd = not determined.

8.3. Schwann Cell Proliferation

Schwann cell proliferation has been studied in detail in the rat. In vivo, maximal proliferation rates of rat Schwann cells occur 2 days before birth (Stewart et al., 1993) whereas in mice, maximal division occurs just after birth (Brown and Asbury, 1981). In the latter case, however, embryonic nerves were not examined.

In vitro, rat Schwann cells divide very slowly in serum containing medium whereas under similar conditions mouse Schwann cells divide rapidly (Krikorian et al., 1982). Where comparisons can be made, the mitogenic requirements for rat and mouse cells appear to be similar, for example both species of Schwann cell divide in response to fibroblast growth factor 2 (FGF2) (Krikorian et al., 1982; Stewart et al., 1991). However, human Schwann cells from adult nerves respond to a more restricted range of mitogens (Morrisey et al., 1991). In these cells high rates of proliferation can be achieved only with a combination of glial growth factor (GGF) and cAMP elevating agents (Morrisey et al., 1991; Rutkowski et al., 1992). These agents are also mitogenic for rat Schwann cells (Lemke, 1990; Stewart et al., 1991).

Fetal human Schwann cells (8–10 wk of gestation) respond to a greater range of mitogens than adult cells (Yong et al., 1988). GGF, FGF2, platelet derived growth factor (PDGF), and nerve growth factor (NGF) are mitogenic for these cells in the presence of serum. Again, all of these factors except NGF are mitogenic for rat Schwann cells. Although few data exist it may be that human Schwann cells, unlike rat Schwann cells (Dong, Stewart, Jessen, and Mirsky, unpublished observation) decrease their responsiveness to mitogens with increasing developmental age.

8.4. Schwann Cell Differentiation In Vitro

It is well documented that differentiation of Schwann cells that is in vivo driven by the axon, is, at least in the rat, mimicked in vitro by agents that elevate intracellular cAMP levels (Jessen and Mirsky, 1991). In particular, cAMP elevating agents induce rat Schwann cells to express myelin proteins and the lipids 04 and galactocerebroside (GalC). This is likely to mean one of two things: either axonal differentiation signals act, in part, by elevating cAMP in Schwann cells, or, alternatively, pathways activated by pharmacological elevation of cAMP in vitro converge at some point with (and therefore mimic) the signalling pathways activated by axonal signals. In contrast to these observations on rat cells, preliminary investigations using fetal and adult human Schwann cells report that cAMP does not induce GalC expression (Pleasure et al., 1985; Kim et al., 1989). Clearly, in order to establish whether cAMP, or cAMP activated pathways, have a general role to play in Schwann cell development studies need to be carried out in a variety of mammalian species.

9. Conclusions

Despite being a major component of the peripheral nervous system in a large number of phyla, detailed knowledge about peripheral glia is very patchy, often making comparisons difficult. Where they can be made, it appears that many of the functions of peripheral glia are conserved amongst morphologically and phylogenetically disparate organisms. However, many more studies are needed in this area. Even within the vertebrates little information regarding the cell biology of Schwann cells in orders other than rodents exists. It has been shown in other cell systems that the molecular machinery involved in developmental events is well conserved between species (Wray, 1994), highlighting the importance in continuing phylogenetic studies.

Acknowledgments

This work was supported by a grant from the Wellcome Trust and the Medical Research Council of Great Britain.

References

Assouline JG Pantazis NJ (1989) Localization of the nerve growth factor receptor on fetal human Schwann cells in culture. Exp Cell Res 182: 499–512.
Barres BA Chun LLY Corey DP (1990) Ion channels in vertebrate glia. Ann Rev Neurosci 13:441–474.
Bergsteinsdóttir K, Kingston AE Mirsky R Jessen KR (1991) Rat Schwann cells produced interleukin-1. J Neuroimmunol 34:15–23.
Bergsteinsdóttir K Kingston AE Jessen KR (1992) Rat Schwann cells can be induced to express major histocompatibility complex class II molecules in vivo. J Neurocytol 21:382–390.
Bevan S (1990) Ion channels and neurotransmitter receptors in glia. Semin Neurosci 2:467–481.
Bevan S Katz B Miledi R (1973) Induced transmitter release from Schwann cells and its suppression by actinomycin D. Nature 241:85–86.
Bhattacharyya A Frank E Ratner N Brackenbury R (1991) Po is an early marker of the Schwann cell lineage in chickens. Neuron 7:831–844.
Bianchini D De Martini I Cadoni A Zicca A Tabaton M Schenone A Anfosso S Akkad Wattar AS Zaccheo D Mancardi GL (1992) GFAP expression of human Schwann cells in tissue culture. Brain Res 570:209–217.
Bittner GD (1991) Long term survival of anucleate axons and its implications for nerve regeneration TINS 14:188–193.
Black JA Waxman S (1990) Ion channel organization of the myelinated fiber. TINS 13:48–54.
Blanco RE (1988) Glial cells in the peripheral nerve of the cockroach *Periplaneta americana*. Tissue and Cell 20:771–782.
Blanco RE Lane NJ (1990) Changes in intercellular junctions during peripheral nerve regeneration in insects. J Neurocytol 19:873–882.
Bray GM Rasminsky M Aguayo AJ (1981) Interactions between axons and their sheath cells. Ann Rev Neurosci 4:127–162.
Brown ER Abbott NJ (1993) The ultrastructure and permeability of the Schwann cell layer surrounding the giant axon of the squid *Alloteuthis subulata*. J Neurocytol 22:283–298.

Brown MJ Asbury AK (1981) Schwann cell proliferation in the postnatal mouse:timing and topography. Exp Neurol 74:170–186.

Buchheit TE Tytell M (1992) Transfer of molecules from glia to axon in the squid may be mediated by glial vesicles. J Neurobiology 23:217–230.

Bullock TH Horridge GA (1965) Structure and function in the nervous system of invertebrates. W. H. Freeman and Co., San Francisco.

Bullock TH Moore JK Fields D (1984) Evolution of myelin sheaths: both lamprey and hagfish lack myelin. Neurosci Lett 48:145–148.

Bunge MB Wood PM Tynan LB Bates ML Sanes JR (1989) Perineurium originates from fibroblasts: demonstration in vitro with a retroviral marker. Science 243:229–231.

Chiu SY (1987) Sodium channels in axon-associated Schwann cells from adult rabbits. J Physiol 386:181–203.

Cobb JLS (1991) Enigmas of the echinoderm nervous system, in Phylogenetics: The Theory and Practice of Phylogenetic Systematics (Wiley EO, ed.), Wiley, New York, pp. 329–337.

Cravioto H (1965) The role of Schwann cells in the development of human peripheral nerves. J Ultrastuct Res 12:634–651.

Curtis R Stewart HJS Hall SM Wilkin GP Mirsky R Jessen KR (1992) GAP-43 is expressed by nonmyelin-forming Schwann cells of the peripheral nervous system. J Cell Biol 116:1455–1464.

Evans PD (1974) An autoradiographical study of the localization of the uptake of glutamate by the peripheral nerves of the crab *Carcinus maenas*. J Cell Sci 14:315–367.

Gamble HJ Breathnach AS (1965) An electron-microscopic study of foetal human peripheral nerves. J Anat 99:573–584.

Gamble HJ (1966) Further electron microscopic studies of human foetal peripheral nerves. J Anat 100:487–502.

Gardner-Medwin AR (1983) A study of the mechanisms by which potassium moves through brain tissue in the rat. J Physiol 335:353–374.

Gardner-Medwin AR Coles JA Tsacopoulos M (1981) Clearance of extracellular potassium: evidence for spatial buffering by glial cells in the retina of the drone. Brain Res 209:452–457.

Geren BB (1954) The formation from the Schwann cell surface of myelin in the peripheral nerves of chick embryos. Exp Cell Res 7:558–562.

Giangrande A (1994) Glia in the fly wing are clonally related to epithelial cells and use the nerve as a pathway for migration. Development 120:523–534.

Golding JP Tonge DA (1993) Expression of GAP-43 in normal and regenerating nerves in the frog. Neuroscience 52:415–426.

Goldstein RS Weiss KR Schwartz JH (1982) Intraneuronal injection of horseradish peroxidase labels glial cells associated with the axons of the giant metacerebral neuron of Aplysia. J Neuroscience 2:1567–1577.

Gray PTA Ritchie JM (1985) Ion channels in Schwann and glial cells. TINS 8:411–415.

Hammer JA O'Shannessy DJ De Leon M Gould R Zand D Daune G Quarles RH (1993) Immunoreactivity of PMP-22, Po and other 19 to 28 kDa glycoproteins in peripheral nerve myelin of mammals and fish with HNK1 and related antibodies. J Neurosci Res 35:546–558.

Heuser JE Doggenweiler CF (1966) The fine structural organization of nerve fiberS sheathS and glial cells in the prawn *Palaemonetes vulgaris*. J Cell Biol 30:381–403.

Jaubarteau MO Jacque C Preud'homme JL Vallat JM Baumann N (1992) Human Schwann cells in culture: characterization and reactivity with human anti-sulfated glucuronyl glycolipid monoclonal IgM antibodies. Neurosci Lett 139:161–164.

Jeserich G Waehneldt TV (1986) Bony fish myelin: evidence for common major structural glycoproteins in central and peripheral myelin of trout. J Neurochem 46:525–533.

Jessen KR Mirsky R Dennison ME Burnstock G (1979) GABA may be a neurotransmitter in the vertebrate peripheral nervous system. Nature 281:71–74.

Jessen KR Morgan LM Stewart HJS Mirsky R (1990) Three markers of adult nonmyelin forming Schwann cellS 217c (Ran 1), A5E3 and GFAP: development and regulation by neuron-Schwann cell interactions. Development 109: 91–103.

Jessen KR Mirsky R (1991) Schwann cell precursors and their development. Glia 4:185–194.

Jessen KR Brennan A Morgan L Mirsky R Kent A Hashimoto Y Gavrilovic J (1994) The Schwann cell precursor and its fate: a study of cell death and differentiation during gliogenesis in rat embryonic nerves. Neuron 12:1–20 .

Katz B Miledi R (1959) Spontaneous subthreshold activity and denervating amphibian end-plates J Physiol 146:44–45P.

Kim SU Yong VW Watabe K, Shin DH (1989) Human fetal Schwann cells in culture: phenotypic expressions and proliferative capability. J Neurosci Res 22:50–59.

Konishi T (1990) Dye coupling between mouse Schwann cells. Brain Res 508:85–92.

Krikorian D Manthorpe M Varon S (1982) Purified mouse Schwann cells: mitogenic effects of fetal calf serum and fibroblast growth factor. Dev Neurosci 5:77–91.

Kuffler SW (1967) Neuroglial cells: physiological properties and potassium mediated effect of neuronal activity on the glial membrane potential. Proc R Soc B 168:1–21.

Lane NJ 1981 Invertebrate neuroglia-junctional structure and development. J Exp Biol 95:7–33.

Lasek RJ Gainer H Przybylski RJ (1974) Transfer of newly synthesized proteins from Schwann cells to the squid giant axon. Proc Natl Acad Sci USA 71:1118–1192.

Le Douarin NM Smith J (1988) Development of the peripheral nervous system from the neural crest. Ann Rev Cell Biol 4:375–404.

Lemke G (1988) Unwrapping the genes of myelin. Neuron 1:533–543.

Lemke G (1990) Glial growth factors. Semin Neurosci 2:437–443.

Maddrell SHP Treherne JE (1967) The ultrastucture of the perineurium in two insect species, *Carausius morosus* and *Periplaneta americana*. J Cell Sci 2:119–128.

Miyazaki T Tasaka J Sakai S Hashiguchi T Padjen AL Tosaka T (1994) Characteristics of Na⁺ current in Schwann cells cultured from frog sciatic nerve. Glia 10:276–285.

Morrisey TK, Kleitman N Bunge RP (1991) Isolation and junctional characterization of Schwann cells derived from adult peripheral nerves. J Neuroscience 11:2433–2442.

Orkand P Kravitz EA (1971) Localization of the sites of γ-amino butyric acid (GABA) uptake in lobster nerve-muscle preparations. J Cell Biol 49:75–89.

Pentreath VW (1987) Functions of invertebrate glia, in Nervous Systems in Invertebrates (Ali M, ed.), Nato ASI series A, vol. 141, Plenum, New York, pp. 61–103.

Pentreath VW (1989) Invertebrate glia cells. Comp Biochem Physiol 93A:77–83.

Perry V H Brown MC (1992) Role of macrophages in peripheral nerve degeneration and repair. BioEssays 14:401–406.

Peters A (1960) The structure of the peripheral nerves of the lamprey *(Lampetra fluviatilis)*. J Ultr Res 4:349–359.

Peters A Muir AR (1959) The relationship between axons and Schwann cells during development of peripheral nerve fibres. Quart J Exptl Physiol 44:117–130.

Pleasure D Kreider B Shuman S Sobue S (1985) Tissue culture studies of Schwann cell proliferation and differentiation. Dev Neurosci 7:364–373.

Radojcic T Pentreath VW (1979) Invertebrate glia. Prog Neurobiol 12:115–179.

Romer A (1970) The Vertebrate Body, 4th ed. W. B. Saunders, Philadelphia.

Roots BI (1978) A phylogenetic approach to the anatomy of glia, in Dynamic Properties of Glia Cells (Schoffeniels E, ed.), Pergamon, New York, pp. 45–54.

Roots BI (1986) Phylogenetic development of astrocytes, in Astrocytes, vol. 1 (Fedoroff, S, Vernadakis, A, eds.), Academic, New York, pp. 1–34.

Roots BI (1993) The evolution of myelin, in Advances in Neural Science, vol. 1 (Nalhotra S, ed.) JAI, Greenwich, CT, pp. 187–213.

Rotshenker S Saada A Aamar S Reichert F (1992) Peripheral nerve injury induces Schwann cells to express two macrophage phenotypes: phagocytosis galactose-specific lectin MAC-2. J Neuroscience (abst) 18 (6084) 1448.

Rutkowski JL Tennekoon GI McGillicuddy JE (1992) Selective culture of mitotically active human Schwann cells from adult sural nerves. Ann Neurol 31:580–586.

Saavedra RA Fors L Aebersold RH Arden B Horvath S Sanders J Hood L (1989) The myelin proteins of the shark brain are similar to the myelin proteins of the mammalian peripheral nervous system. J Mol Evol 29:149–156.

Scarpini E Meola G Baron P Beretta S Velicogna ME Scarlato G (1986) S-100 protein and laminin: immunocytochemical markers for human Schwann cells in vitro. Exp Neurol 93:77–883.

Scarpini E Ross AH Rosen JL Brown MJ Rostami A Koprowski H Lisak, RP (1988) Expression of nerve growth factor receptor during human peripheral nerve development. Dev Biol 125:310–310.

Schon F Kelly JS (1974) Autoradiographic localization of ³H-GABA and ³H-Glutamate over satellite glial cells. Brain Res 86:499–503.

Shearman JD Franks AJ (1987) S-100 protein in Schwann cells of the developing human peripheral nerve. Cell Tiss Res 249:459–463.

Stewart HJS Eccleston PA Jessen KR Mirsky R (1991) Interactions between cAMP elevation identified growth factors and serum components in regulating Schwann cell growth. J Neurosci Res 30:346–352.

Stewart HJS Morgan L Jessen KR Mirsky R (1993) The relationship between DNA synthesis embryonic development and myelination in the rat sciatic nerve. Eur J Neurosci 5:1136–1146.

TakeI K Uyemura K (1993) Expression of Po-like glycoprotein in central nervous system myelin of amphibians (*Ambystoma mexicanus, Xenopus laevis and Rana catesbeiana*). Comp Biochem Biophysiol 106B:873–882.

Tytell M Greenberg SG Lasek RJ (1986) Heat shock-like protein is transferred from glia to axon. Brain Res 363:161–164.

Uyemura K, Horie K, Kitamura K, Suzuki M Uehara S (1979) Developmental changes of myelin proteins in the chick peripheral nerve. 32:779–788.

Viancour TA Bittner GD Ballinger ML (1981) Selective transfer of lucifer yellow CH from axoplasm to adaxonal glia. Nature 293:65–67.

Villegas J (1981) Axon/Schwann cell relationships in the giant nerve fiber of the squid. J Exp Biol 95:135–151.

Villegas J (1984) Axon-Schwann cell relationships. Curr Topics Membr Transport 22:547–571.

Webster HD (1971) The geometry of peripheral myelin sheaths during their formation and growth in rat sciatic nerves. J Cell Biol 48:348–367.

Weston JA (1991) Sequential segregation and fate of developmentally restricted intermediate cell populations in the neural crest lineage. Curr Topics Dev Biol 25:133–153.

Wray G (1994) Developmental Evolution: new paradigms and paradoxes. Dev Genet. 15:1–6.

Yan Q Johnson EM (1987) A quantitative study of the developmental expression of the nerve growth factor (NGF) receptor in rats. Dev Biol 121:139–148.

Yong VW Kim SU Kim MW Shin DH (1988) Growth factors for human glial cells in culture. Glia 1:113–123.

Zimmerman A Suter A (1983) Nerve growth factor receptors on glial cells: cell-cell interaction between neurons and Schwann cells in cultures of chick sensory ganglia. EMBO J 2:879–885.

Glial Interactions with Neurons During *Drosophila* Development

John R. Fredieu and Anthony Mahowald

1. Introduction

Neurons present in the *Drosophila* embryo do not exist in an exclusively neuronal environment. Cells that lack neuronal characteristics are interspersed with neurons and their axons throughout the segmental ganglia and the neuropil in the central nervous system (CNS) and in association with neurons and axons of the peripheral nervous system (PNS) (Scharrer, 1939; Poulson, 1950; Smith and Treherne, 1963; Springer and Rutschky, 1968; Vanhems, 1985; Bastiani and Goodman, 1986; Hoyle, 1986; Meyer et al., 1987; Fredieu and Mahowald, 1989; Jacobs et al., 1989; Jacobs and Goodman, 1989a). In fact, the majority of cells within the CNS of mature insects may be glial in nature (Becker, 1965). Poulson suggested that a number of cells in the ventral median cord of the *Drosophila* embryo are glia and that they provide support for the developing neurons and their axons (Poulson, 1950). He postulated that these glial cells function in providing the necessary components and support for the proper elaboration of the central neuropil and peripheral axon projections, functions that correspond to those proposed in the earliest studies of arthropod neural development (Nusbaum, 1883; Wheeler, 1893; Ramon y Cajal and Sanchez, 1915).

More recently, defined groups of cells within the ventral midline of the *Drosophila* embryo have been identified as glia based on specific morphological criteria, antigenic determinants, and lineage analysis (Fredieu and Mahowald, 1989; Jacobs et al., 1989;

From: *Neuron-Glia Interrelations During Phylogeny: I. Phylogeny and Ontogeny of Glial Cells*
A. Vernadakis and B. Roots, Eds. Humana Press Inc., Totowa, NJ

Jacobs and Goodman, 1989a; Klambt and Goodman, 1991; Nelson and Laughton, 1993). In these same studies, other glial cells have been described which are associated with the longitudinal axon tracts of the CNS and with axon tracts of the PNS. In addition, the entire CNS is ensheathed by a layer of non-neuronal cells, called the perineurium, although the classification of these cells as glia is controversial (Smith and Treherne, 1963; Sohal et al., 1972; Hoyle, 1986; Bunge et al., 1989; Smith et al., 1991; Edwards et al., 1993). All of these cells have been identified as glia by means of several morphological characteristics, including shape (large multipolar nuclei, multiple cellular processes), cytoplasmic structures (rER in cytoplasm of processes, glycogen bodies), and neurotransmitter uptake properties (glutamate, GABA). Individual cells are also classified as glia by the lack of antigenic determinants specific for neurons and the expression of other antigenic epitopes that are not found in neurons (Jan and Jan, 1982; Fredieu and Mahowald, 1989; Jacobs and Goodman, 1989a; Ebens et al., 1993; Nelson and Laughton, 1993). In addition, using mutational analysis, the cells classified as glia in *Drosophila* by the above criteria have been found to possess comparable functional roles with those of vertebrate glia (Nambu et al., 1990; Rothberg et al., 1990; Klambt et al., 1991).

Glial cells have been ascribed a variety of functions within the developing and mature nervous system of many vertebrate and invertebrate species. These functions include the ionic compartmentalization of neurons and axons (Coles and Orkland, 1983; Remahl and Hildebrand, 1990), management of the microenvironment in the form of release and uptake of factors and nutrients (Faeder and Salpeter, 1970; Reynolds and Herschkowitz, 1986; Beadle et al., 1987), injury and repair (Aguayo et al., 1981; Tytell et al., 1986; Treherne et al., 1987), and the provision of an adequate substrate for the sprouting and guidance of neurites during development and regeneration (Noble et al., 1984; Fallon, 1985; Bastiani and Goodman, 1986; Jacobs and Goodman, 1989a,b). Glial cells also exchange information with neurons and other cells through the use of direct cellular contact and the use of secreted factors (Bunge et al., 1982; Tytell et al., 1986; Berkley and Contos, 1987). All of these functions are used during embryonic development to establish the proper structure of the segmental ganglia, the proper elaboration of the CNS and PNS axon pathways, and an environment conducive to neuronal function.

The role of the glial cell in the development of the embryo, as well as its role in the maintenance of embryonic cellular environments, advance the proposed biological functions of glia from a support cell for neurons to a predominant role in the development and function of the complex nervous system. The glial cell in the *Drosophila* embryo is crucial to the functional architecture of the mature CNS through its role in the guidance of neurites along their proper pathways, boundary formation that delimits the extent of neurite elongation, and morphogenic movements that define the mature structure of the embryonic CNS. Contrary to the importance of glial cells during development, little is known about their developmental and phenotypic heterogeneity relative to the information available concerning identified neurons in the CNS and PNS of the *Drosophila* embryo.

2. Subpopulations of Drosophila *Glia*

Glial cells in the *Drosophila* embryo can be divided into subpopulations dependent on their function and lineage (Fig. 1). Most of these subpopulations share a lack of expression of neuronal antigens, but differ greatly in their cellular structure, proposed biological function, and embryonic lineage. Although the identification of all of the specific glial cells in the *Drosophila* embryo is most likely incomplete, the available information that has been generated over the past several years has facilitated the classification of all of the identified glial cells into specific subpopulations based on functional or lineage relationships with each other and the surrounding neurons.

2.1. Commissural and Longitudinal Glia

The ventral midline of the *Drosophila* embryo contains glial cells that are thought to function in the development of both the segmental commissures and the longitudinal connectives of the mid- to late embryo. The anterior and posterior commissures are composed of groups of fasciculated axons that originate from neurons on one side of the midline and cross to the contralateral side of the embryo (Fig. 1). Ensheathing these axon fascicles in the late embryo are three pairs of ventral midline glia in each segment that have been identified by specific antigenic determinants, morphological analysis, gene expression, and enhancer-specific

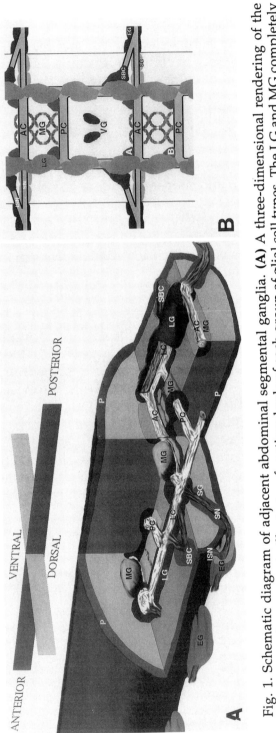

Fig. 1. Schematic diagram of adjacent abdominal segmental ganglia. **(A)** A three-dimensional rendering of the same abdominal segments to illustrate a functional role of each group of glial cell types. The LG and MG completely ensheath the central neuropil, whereas other glial cells ensheath the axon fascicles of the segmental nerve (SN) and the intersegmental nerve (ISN) of the peripheral nervous system. The glial cells that are interspersed throughout the neuronal cell bodies are not illustrated here. **(B)** Segments of the neuropil are associated with several identified groups of glial cells. Longitudinal glia (LG) and midline glia (MG) are subdivided into anterior, middle, and posterior components. Each component of the LG and MG maintain their relationship with the longitudinal and posterior components. Each component of the LG and MG maintain their relationship with the longitudinal and posterior commissures (PC), respectively. Abbreviations: AC, anterior connectives (LC) and the anterior (AC) or posterior commissures (PC), respectively. Abbreviations: AC, anterior connectives; EG, exit glia; ISN, intersegmental nerve; LC, longitudinal connective; LG, longitudinal glia; MG, midline glia; P, perineurium; PC, posterior commissure; SBC, segmental boundary cell(s); SG, segmental nerve glia; SN, segmental nerve.

expression of reporter genes (Fredieu and Mahowald, 1989; Jacobs and Goodman, 1989a,b; Nambu et al., 1990; Rothberg et al., 1990; Klambt et al., 1991; Smith and O'Kane, 1991; Hartenstein and Jan, 1992). The midline glia segregate the axons that course through each commissure by ensheathing either individual or groups of axons by the mid- to late embryo. The ensheathment of specific axons or the inclusion of a specific axon into a segregated group may depend on factors such as location of the target cell for the neurite or specific antigenic determinants expressed on the axon's surface. The anterior and posterior commissures depend on these midline glia for the guidance of their original pioneer axons that cross the ventral midline to establish the commissures. Subsequent neurites may depend on specific pioneer axons as a pathway or guidance substrate. In this way, individual fascicles are established by specific directional cues recognized by the later neurites.

The longitudinal connectives are bundles of axon fascicles connecting the segmental ganglia between adjacent segments. In the same manner, as the midline glia provide an axon guidance substrate and ensheath the segmental commissures, the axons traveling through the longitudinal connectives depend on three pairs of longitudinal glia on either side of the ventral midline in each segment for their initial pathway guidance and subsequent ensheathment (Jacobs et al., 1989).

In a series of comprehensive studies by Goodman and colleagues, the midline glia and the longitudinal glia have been shown to be critical for the proper formation and structure of the segmental commissures and the longitudinal connectives respectively (Jacobs et al., 1989; Jacobs and Goodman, 1989a,b; Klambt et al., 1991; Klambt and Goodman, 1991b). Using morphological and genetic techniques, these studies demonstrated a phasic nature to the interactions between glia and neurons of the *Drosophila* CNS. Although being important in establishing the axon pathway, once the pathway is pioneered by the initial neurites, the glial cells begin to ensheath the axon fascicles and alter their relationship with the segmental ganglia by segregating and repositioning the commissure to ensure their proper alignment in the late embryo.

In addition to these main classes of glial cells, other populations of glial cells have been identified in the CNS of the *Drosophila* embryo. The "A" and "B" glia, the dorsal roof cells, and the intersegmental glia 1 and 2 (may be homologous to the segmental boundary cells in Fig. 1) have been identified by the use of enhancer-specific and transcription factor-specific expression of reporter genes during embryonic development (Klambt and Goodman, 1991a; Nelson and Laughton, 1993). Other glial cells are associated exclusively with the neuronal cell bodies of the segmental ganglia and of the brain lobes and are separate from the neuropil glia. Although little is known of the functions of these subpopulations of glial cells, their temporal and spatial appearance during significant periods of neurogenesis and axonogenesis provide clues to their role in development. For example, whereas the glia associated with the embryonic neuropil may provide a substrate for growth cone guidance and may ensheath the axon fascicles, glial cells associated with the neuronal cell bodies may condition the extracellular microenvironment to provide various developmental cues for adjacent neurons (i.e., neurotrophic factors or factors influencing cellular proliferation). In addition, these same glial cells may provide mechanical cues for the neuron to orient itself and the position of its neurites and thus, structurally polarize the cell soma.

2.2. Peripheral Glia

The PNS of the *Drosophila* embryo also contains cells that have been identified as glia through similar techniques as used to characterize the CNS glia (Fredieu and Mahowald, 1989; Klambt and Goodman, 1991a; Nelson and Laughton, 1993). The peripheral glia can be subdivided by their location, lineage, and their relationship to specific peripheral neurons (Fig. 1). Glial cells and other non-neuronal cells are associated with the ventral, lateral, and dorsal groups of sensory organs and their axons that comprise the segmental and intersegmental peripheral nerves. These cells have been identified as exit glia (EGA, EGM1, EGM2) and are associated with the segmental and intersegmental nerves as they exit the CNS, and with other glia (PG1-3) along the path of the peripheral nerves (Klambt and Goodman, 1991; Nelson and Laughton, 1993). The segmental and intersegmental nerves in the late embryo are ensheathed by these glial cells (Fredieu and Mahowald, 1989).

2.3. Perineurium

The perineurium is a layer of cells that covers the surface of the CNS and secretes a connective tissue sheath termed the neural lamella (Scharrer, 1939; Hoyle, 1986). Underlying the perineurium is a monolayer of glial cells called the barrier cells (Smith et al., 1991 for review; Edwards et al., 1993). This layer of cells is thought to be the functional equivalent of the blood/brain barrier found in vertebrates. There is much controversy as to the classification of the cells of the perineurium as glia owing to their doubtful ectodermal origin (Scharrer, 1939; Bunge et al., 1989; Edwards et al., 1993). The perineurium and the barrier cell layer are detectable using an antibody specific for glial cells (Mab5B12); however, several techniques which detect other populations of glial cells in *Drosophila* have failed to label the barrier cells or the perineurium (Fredieu and Mahowald, 1989; Nelson and Laughton, 1993).

3. Neurogenesis

The complexity of the nervous system in *Drosophila* becomes evident during the initial stages of embryogenesis and increases throughout development until emergence of the adult fly. Initially, the CNS is divided into metameric units along the germ band corresponding to each parasegment. Each parasegment is composed of lateral neuromeres separated by the ventral midline. Neuroblasts begin to delaminate from the neuroectoderm within and lateral to the ventral midline and move centripetally to occupy positions dorsal to the remaining ventral ectoderm (Thomas et al., 1984; Doe and Goodman, 1985a). After neuroblasts have segregated from the ectoderm, they proceed through a series of asymmetric divisions to form ganglion mother cells (Fig. 2A) (Poulson, 1950; Furst and Mahowald, 1985; Hartenstein et al., 1987). Each ganglion mother cell then divides symmetrically into a pair of neurons. One neuroblast is capable of producing 6–20 neurons between 7 and 13 h after fertilization (Furst and Mahowald, 1985; Hartenstein et al., 1987). Daughter neurons derived from an individual neuroblast have been demonstrated to be phenotypically distinct and, at least in some cases, dependent on the birth date of the ganglion mother cell (Doe and Goodman, 1985b; Huff et al., 1989; Doe and Technau, 1993). Phenotypic diversity of sibling neurons from a single ganglion mother cell is less well documented, but studies indicate that sibling neurons are able to express different cell surface glycoconjugates and send out neurites that respond to divergent guidance cues (Thomas et al., 1984; Fredieu and Mahowald, 1994).

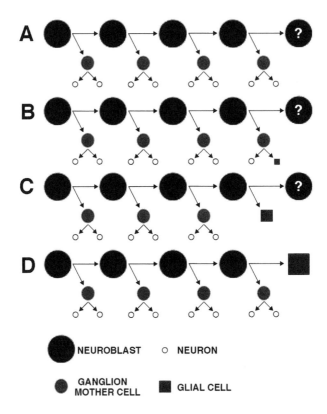

Fig. 2. Division sequences of the neuroblast that may give rise to a glial cell. **(A)** Typical asymmetric division of a neuroblast into several smaller ganglion mother cells. Each ganglion mother cell (GMC) then divides symmetrically into a pair of neurons. Only a few divisions of the neuroblast are depicted in that the neuroblast may divide several more times. A neuroblast as the progenitor of both neurons and glia may produce neurons in the same fashion as in A, but glial cells may be derived from a resultant GMC, as in **B**, or the GMC itself may differentiate into a glial cell, as in **C**. In addition, the terminal division of the neuroblast may produce a GMC and a glial cell as in **D**. Abbreviations: AC, anterior commissure; EG, exit glia; ISN, intersegmental nerve; LC, longitudinal connective; LG, longitudinal glia; MG, midline glia; P, perineurium; PC, posterior commissure; SBC, segmental boundary cell(s); SG, segmental nerve glia; SN, segmental nerve.

During the extensive neuronal proliferation coincident with germ band shortening, neurons within the ventral midline begin to extend neurites with growth cones (Canal and Ferrus, 1986; Ghysen et al., 1986; Jacobs and Goodman, 1989a,b). The initial axonal growth cones (pioneers) emanate from medial and dorsal neurons, cross the ventral midline to the contralateral hemisegmental ganglion and pioneer the anterior and posterior commissures of the embryonic segmental ganglion, as well as, the intersegmental nerve (ISN) (Jacobs and Goodman, 1989a). At the same time, other pioneer growth cones move anteriorly and posteriorly to form the longitudinal connectives. Formation of the complete embryonic central neuropil is dependent on the proper pathway selection of these pioneer growth cones (Thomas et al., 1984; Bastiani et al., 1986; Doe et al., 1986).

The *Drosophila* PNS begins to differentiate during germ band contraction and is structurally complete by the end of dorsal closure (Ghysen et al., 1986; Bodmer et al., 1989). Cells within the ectoderm along the dorso-ventral axis in each segment divide and differentiate into neuronal and non-neuronal components of the PNS (Bodmer and Jan, 1987). The more dorsal components appear first and are quickly followed by lateral and ventral cells. The resultant cell types that comprise the PNS differ dramatically between cephalic, thoracic, and abdominal segments.

Axonal pathways of the PNS are derived by growth cones of pioneer axons moving dorsolaterally from the CNS and peripheral pioneer growth cones moving ventrally from the dorsal sensilla (Canal and Ferrus, 1986; Ghysen et al., 1986; Bodmer and Jan, 1987). These opposing pioneers meet and form the ISN and segmental nerves (SN) of the PNS.

4. Gliogenesis

The earliest glial progenitors in the CNS are two types of glioblasts that differ in location and their tissue source. At the onset of gastrulation, mesodermal cells, which can be identified by their *single-minded* (*sim*) expression, invaginate along the ventral midline (Crews et al., 1988; Thomas et al., 1988; Jacobs et al., 1989). As the cells along each lateral edge of the *sim*-expressing region fuse at the midline, the mesectoderm becomes distinguishable as a set of cells stretching vertically from within the ectoderm through the newly formed mesoderm. A subset of mesoectodermally derived cells in the ventral midline divide once and produce the anterior, middle, and posterior midline glia (Jacobs and Goodman, 1989a; Klambt et al., 1991). These progenitor cells, or midline precursors, express *sim* and initially are detectable after the completion of gastrulation. By the end of germ band elongation, the midline glia are the only cells of the mesectoderm that continue to express *sim*. After axonogenesis is well underway, the anterior (MGA), middle (MGM), and posterior (MGP) midline glial cells realign themselves along the processes of the ventral unpaired median neurons (VUMs) and align the anterior and posterior commissures to their normal positions (Klambt and Goodman, 1991b).

Glial progenitors are located lateral to the ventral midline in each hemisegment of the CNS. These glioblasts divide once, migrate, and then continue to proliferate to produce six longitudinal glia in each neuromere (Jacobs et al., 1989). The progenitors of the longitudinal glia are ectodermally derived and originate from the layer of neuroblasts delaminating from the ventral ectoderm. The longitudinal glia are further distinguishable from the midline glia in that they express the gene *fushi tarazu*, an ectodermally expressed gene, and do not express the meso- or mesectodermal marker, *sim* (Doe et al., 1988; Jacobs et al., 1989). As the longitudinal glia align themselves ventral to the presumptive position of the longitudinal connectives, the initial pioneer neurites that frame the path of these large axon fascicles use these glial cells as a support and guidance substrate.

In the last few years, several reports have strongly suggested the existence of bipotential progenitor cells in the *Drosophila* embryo that are able to divide and produce both neurons and glia in the CNS and in the PNS (Bodmer and Jan, 1987; Fredieu and Mahowald, 1989; Becker and Technau, 1990; Udolph et al., 1993). Using single-cell labeling techniques, the lineage of one neuroglioblast of the CNS, the NB 1-1, has been shown to produce ganglion mother cells, some of which further divide into both neurons and a set of glial cells (Udolph et al., 1993). These glial cells were subsequently found to be identical to the "A" and "B" glial cells described earlier. The NB 1-1 neuroblast does not produce glial cells in the thoracic segmental ganglia but only in abdominal segments, suggesting either a segmental specificity in neuroglioblast determination, or the respecification of the cells destined to become glial cells in the thoracic segments. The homeotic transformation of the NB 1-1 cell in abdominal segments to a thoracic segment lineage is maintained when the mutant cells are transplanted to abdominal segments of the wild type embryo, suggesting a cell autonomous requirement for the homeotic gene products to specify the production of glial cells by the NB1-1 cell (Doe and Technau, 1993).

There are additional neuroglioblasts that have been identified in the CNS and in the PNS. The VUM cell produces several neurons within the ventral midline and a variable number of glial cells (Klambt and Goodman, 1991a). The VUM cell is derived from the mesectoderm, as are the midline glia and, as described previously, the midline glial cells utilize the processes of the VUM progeny as migratory substrates. Other neuroglioblasts exist in the periphery and divide to produce specialized sensory neurons, glial cells, and support cells of the PNS (Bodmer et al., 1989). For example, the bipolar dendrite, which lies just dorsal to the lateral sensilla, divides once to produce a neuron and a glial cell (Bodmer et al., 1989; Nelson and Laughton, 1993). Other progenitor cells in the PNS produce both neurons and non-neuronal support cells.

In addition to specific glioblasts and neuroglioblasts found in the *Drosophila* embryo, some neuronal cell clusters in vitro, derived from dissociated embryonic neuroblasts maintained in culture, contain cells that express a glial-specific surface antigen (antigen to Mab5B12) (Fredieu and Mahowald, 1989). When neuroblasts are isolated from *Drosophila* gastrulae, enriched by cell elutriation techniques, and allowed to develop for 24 h in culture, they produce ganglion mother cells that divide into sibling neurons in vitro with similar characteristics as those found in vivo (Huff et al., 1989). Approximately 40% of the resultant neuronal clusters in culture will produce a Mab5B12-immunoreactive glial cells suggesting that individual progenitors may be capable of producing progeny that express either neuronal or glial phenotypes (Fredieu and Mahowald, 1989). Also found in the differentiated neuroblast cultures are groups of Mab5B12-immunoreactive cells that are presumably the progeny of coisolated glioblasts. These progenitors may correspond to the midline glioblasts that give rise to the midline and longitudinal glia in the embryonic CNS.

The culturing of *Drosophila* cells in media containing hydroxyurea (HU) has been demonstrated to inhibit cell division. Neuroblasts treated with 2mM HU will continue to differentiate in the absence of DNA synthesis, but the normal series of cell divisions that produce the ganglion mother cells is truncated at the time of HU addition (Furst and Mahowald, 1985; Huff et al., 1989). Neuronal clusters derived from neuroblasts treated prior to 3 h after plating (equivalently 7 h old) contain few associated glial or Mab5B12-IR cells, whereas the percentage of glia-containing neuronal clusters derived from neuroblasts treated with HU after longer periods in culture are comparable to untreated cultures (unpublished observations). The neuronal clusters that contain an Mab5B12-IR cell when treated with hydroxyurea at any time prior to two neuroblast divisions may be derived from either a population of cells that have divided during the culture preparation or may be the population of neuronal clusters that obtain a Mab5B12-IR cell through adhesion or other associative mechanisms before the cells attach to the substrate.

The glial cell progeny of a cultured neuroglioblast could arise from distinct progenitor division sequences (Fig. 2). First, either one daughter cell of a ganglion mother cell (the direct progeny of the neuroblast) may be glial, whereas its sibling acquires a neuronal phenotype, or both sibling progeny of the ganglion mother cell may be glial. Second, the product of one of the asymmetric divisions of the neuroblast may terminally differentiate directly into a glial cell rather than into a ganglion mother cell progenitor. Finally, the final asymmetric division of the neuroblast may give rise to ganglion mother cell and a glial precursor. In *Drosophila*, the result of the final division of a neuroblast is not clear, but it is unlikely that the terminal differentiation of a neuroblast to a glial cell is the predominate source of neuroblast-derived glial cells. The observations that a ganglion mother cell may divide into a neuron and a glial cell as in the NB1-1 lineage and the results from truncation of neuroblast divisions suggest that neuroblasts also produce glial cells during the early stages of their division series. The derivation of a glial cell from the specific division of a neuroblast may provide a local interaction between neurons and glia serving to align or segregate cells within a clonal array of neurons early in the development of the neuropil and the ganglia. In addition, glial cells in *Drosophila* have been implicated in the regulation of neuroblast proliferation suggesting an intrinsic role of the daughter glia in a cluster of cells derived from an individual neuroblast (Ebens et al., 1993).

The percentage of neuronal clusters in culture found to express Mab5B12 immunoreactivity is not reflected in the intact embryo (Fredieu and Mahowald, 1989). Several explanations for these observations may apply. First, the mechanical dissociation of the neuroectoderm during the culture procedure may invoke a neurogenic response owing to the lack of suppressive signals generated by cellular contact between cells destined for either the neural or epidermal lineage. Mutations in the neurogenic genes *Notch, mastermind,* and *big brain* result in the overproduction of glioblasts and neuroblasts in vivo (Jacobs and Goodman, 1989a). Second, a number of thoracic neuroblasts, and possibly glioblasts, do not divide until larval stages (Truman and Bate, 1988). These larval neuroblasts may be persistent embryonic neuroblasts and may be present in cultures derived from embryos. If a number of these persistent neuroblasts are actually neuroglioblasts, then an increase in the number of glial cells in culture would be apparent. Third, glioblasts that produce the longitudinal glia in the embryo migrate a short distance before dividing again (Jacobs et al., 1989). This migration and, possibly, the dispersal of the glioblast progeny in culture, may artificially inflate the percentage of neuronal clusters that contain a Mab5B12 immunoreactive cell.

The perineurium consists of Mab5B12 immunoreactive cells at later embryonic stages and it was initially considered as a source of the large numbers of cultured neuronal clusters containing a Mab5B12 immunoreactive cell (Fredieu and Mahowald, 1989). Recent studies, however, suggest that the cells comprising the perineurium are of mesodermal origin (Edwards et al., 1993). *Drosophila* embryos that are homozygous for *twist* fail to develop mesodermal tissues and also lack a perineurium and neural lamella. Although the possibility exists that the cells that produce the perineurium fail to develop because of the lack of a mesodermal interaction with the ectoderm, it is more likely that the perineurium is produced directly from cells of mesodermal origin. Therefore, the cells of the perineurium could not account for the sibling glial cell among neurons derived from a isolated neuroblast. However, the barrier cells, which also may be Mab 5B12 immunoreactive, cover the entire surface of the CNS. These cells lay beneath the perineurium and may arise locally from individual neuroblasts that are dispersed over the surface of the CNS.

The embryonic glioblasts and neuroglioblasts discussed here do not encompass all of the glial cells present in the *Drosophila* embryo. Glial cells that invest the supraesophageal and the subesophageal commissures of the brain and the antenno-maxillary complex, are detectable with most of the methods used to visualize the other glial cells (Fredieu and Mahowald, 1989; Nelson and Laughton, 1993).

5. Axonogenesis

Drosophila has received much attention as a model in the investigation of axon pathway formation through study of embryonic and postembryonic neural development (Thomas et al., 1984; Blair and Palka, 1985; Blair et al., 1987; Thomas et al., 1988). Many studies have relied on individual neuronal identity to investigate cellular determination and pattern formation in neuromere development (Dambly-Chaudière and Ghysen, 1986; Ghysen et al., 1986; Bodmer and Jan, 1987). Perhaps the most comprehensive investigations have focused on specific molecules present on the surface of these identified neurons that may influence the proper routing of CNS axons (Bastiani et al., 1987). Patel et al., 1987; Seeger et al., 1993).

The axon begins as a neurite that sprouts from the neural cell soma and possesses an actively motile growth cone at its distal end (Harrison, 1910; Yamada et al., 1971). The growth cone advances through the embryonic tissue in a directed fashion by the use of plasmalemmal filopodia. The filopodia sample much of the surrounding cell surfaces and the extracellular matrix for adhesive, trophic, and instructive signals. The pathway that a growth cone will take depends on the nature of these epigenetic signals and their effects on the behavior of individual filopodia. The growth cone enables a neurite to extend great distances traversing areas of diverse cellular and extracellular components. Specific spatial and temporal points along the growth cone's path present critical vectoral alternatives as to the location and nature of the final target on which the axon will synapse. The growth cone's commitment to a specific direction is made in response to available signals that are interpretable by the growth cone. The pathway substrate that contains these signals and that the growth cone samples is, in many cases, composed of glial cell surfaces.

Neurites in the *Drosophila* embryo are presented with many directional alternatives, beginning at their initial formation and continuing to the formation of a synapse on the target cell. The initial sprouting of a stable neurite by the neuronal cell soma is polar in orientation and the neurite extends in a directed manner. For example, identified midline precursor cells (MP1, dMP2, and vMP2), sprout neurites laterally at about 10 h of development (Stage 13). The neurites from MP1 and dMP2 then extend posteriorly, whereas the neurite of vMP2 extends anteriorly (Thomas et al., 1984). At the same time, the sibling neurons pCC and aCC send neurites anteriorly and to the ISN, respectively. Recently, genetic mutations have been uncovered that alter the signaling mechanisms used in directing neurites across the ventral midline (Seeger et al., 1993). In one example, embryos homozygous for a mutation in the gene, *commissureless* (*comm*), lack most of the anterior and posterior commissures in the CNS. The initial neurites are oriented toward the midline, as are their wild-type homologs. However, they soon turn away from the midline and grow onto aberrant ipsilateral pathways. In some cases, the pathways chosen by the mutant cells are mirror images of the normal contralateral pathway, suggesting that the defect lies in the substrate cell at the ventral midline and not in the neurons ability to interpret the guidance signals or in subsequent growth cone substrates on the contralateral side.

Growth cones require a substrate on which to move (Letourneau, 1975). The substrates can be either an acellular extracellular matrix deposited by specialized cells or the surfaces of neighboring cells. Growth cones of the vMP2 and the pCC initially contact the surface of a longitudinal glial cell (LG5) and move on its surface in the adjacent anterior segment (Jacobs and Goodman, 1989b). The growth cone of the aCC moves posterior-laterally across the LGs to contact the segmental boundary cell (SBC) and pioneer the ISN. In *Drosophila*, a growth cone can, at any given time, sample many different candidate cellular and acellular substrates. A single filopodium of a typical growth cone in *Drosophila* may reach 15 µm, capable of interacting with 50% of the cells bordering the neuropil within a hemisegment, whereas a complete growth cone is capable of sampling the entire length or more of a hemisegment (Jacobs and Goodman, 1989b). Thus, the distinct specificity in the epigenetic cues directing a growth cone's movement must come from the number of distinct or overlapping signals a growth cone may receive at any given time.

The generation of mutants and the screening of existing mutants that result in a phenotypic alteration of either the axon pathway of individual neurons or the overall structure of the central neuropil have revealed much information about the requirements of the growth cone for selecting its appropriate path (Jacobs et al., 1989; Smouse and Perrimon, 1990; Seeger et al., 1993) The structure of the neuropil of the late embryo does not depend on just the presence of the proper substrate, but also on the alignment and position of the substrate cell and the expression of specific cues by the substrate cells at the proper time. For example, not only must the midline glia express the specific cue(s) that will actively influence the pioneering commissural growth cones to move onto their surfaces, they must express these cues at the moment and position that the growth cone requires. If the midline glia are not in their proper positions, defects in the structure or the formation of the anterior and posterior commissures can be observed (Klambt et al., 1992).

After the axon pathways are formed and the general scaffold plan has been established, the embryonic glial cells of the central and peripheral neuropil ensheath and segregate the axon fascicles from the other cells and extracellular components of the embryo (Fredieu and Mahowald, 1989; Jacobs and Goodman, 1989a). The glial cells also segregate groups of axons from others in the anterior and posterior commissures as well as the ISN. The axons comprising these smaller fascicles may share characteristics that are reflected in their origin and destination, or may express surface components that, if left exposed, would interfere with the proper guidance of subsequent growth cones passing through. It is also possible that later growth cones no longer have to depend on the cell surface guidance cues and are merely following the channels produced by the enveloping the glial cells.

6. Summary

This brief chapter presents the development of glia in *Drosophila* and their role in the morphogenesis of the central and peripheral nervous systems. Clearly, glial cells are important in the fly, as in other organisms, for the proper routing or pioneer growth cones as they course through the embryo toward their final targets. In addition, glial cells are important for the positioning and ensheathment of the axon fascicles in the later embryo. Basic anatomical and histological studies have revealed a wealth of information as to the complexity and diversity of neuron–glial interactions during development. Presently, molecular genetic dissection of the multiple roles an individual glial cell performs during development is proving to redefine the glial cell as a major player in the development of the nervous system in *Drosophila* beyond the label as a neuronal support cell.

References

Aquayo AJ David S Bray GM (1981) Influences of the glial environment on the elongation of axons after injury: transplantation studies on adult rodents. J Exp Biol 95:231–240.

Bastiani MJ du Lac S Goodman CS (1986) Guidance of neuronal growth cones in the grasshopper embryo. I. Recognition of a specific axonal pathway by the pCC neuron. J Neurosci 6:3518–3531.

Bastiani MJ Goodman CS (1986) Guidance of neuronal growth cones in the grasshopper embryo. III. Recognition of specific glial pathways. J Neurosci 6:3542–3551.

Bastiani MJ Harrelson AL Snow PM Goodman CS (1987) Expression of fasciclin I and II glycoproteins on subsets of axon pathways during neuronal development in the grasshopper. Cell 48:745–755.

Beadle CA Bermudez I Beadle DJ (1987) Amino acid uptake by neurons and glial cells from embryonic cockroach brain growing in vitro. J Insect Physiol 33:761–768.

Becker HW (1965) The number of neurons, glia, and perineurium cells in an insect ganglion. Experentia 21:719.

Becker T Technau GM (1990) Single cell transplantation reveals interspecific cell communication in *Drosophila* chimeras. Development 109:821–832.

Berkley KJ Contos N (1987) Aglial-neuronal-glial communication system in the mammalian central nervous system. Brain Res 414:49–67.

Blair SS Palka J (1985) Axon guidance in cultured wing disks and disk fragments of *Drosophila*. Dev Biol 10:8411–8419.

Blair SS Murray MA Palka J (1987) The guidance of axons from transplanted neurons through aneural *Drosophila* wings. J Neurosci 7:4165–4175.

Bodmer R Jan YN (1987) Morphological differentiation of the embryonic peripheral neurons in *Drosophila*. Roux's Arch Dev Biol 196:69–77.

Bodmer R Carretto R Jan YN (1989) Neurogenesis of the peripheral nervous system in *Drosophila* embryos: DNA replication patterns and cell lineages. Neuron 3:21–32.

Bovolenta P Dodd J (1990) Guidance of commissural growth cones at the floor plate in embryonic rat spinal cord. Development 109:435–447.

Bunge MB Williams AK Wood PM (1982) Neuron-Schwann cell interactions in basal lamina formation. Dev Biol 92:449–457.

Bunge MB Wood PM Tynan LB Bates ML Sanes JR (1989) Perineurium originates from fibroblasts: demonstration in vitro with a retroviral marker. Science 243: 229–231.

Canal I Ferrus A (1986) The pattern of early neuronal differentiation in *Drosophila* melanogaster. J Neurogenet 3:293–319.

Coles JA Orkland RK (1983) Modification of potassium movement through the retina of the drone (Apis mellifera male) by glial uptake. J Physiol 340:157–74.

Crews ST Thomas J Goodman CS (1988) The *Drosophila single-minded* gene encodes a nuclear protein with sequence similarity to the *per* gene product. Neuron 10:409–426.

Dambly-Chaudière C Ghysen A (1986) The sense organs of *Drosophila* larva and their relation to the embryonic pattern of sensory neurons. Roux's Arch Dev Biol 195:222–228.

Doe CQ Goodman CS (1985a) Early events in insect neurogenesis. I. Developmental and segmental differences in the pattern of neuronal precursor cells. Dev Biol 111:193–205.

Doe CQ Goodman CS (1985b) Early events in insect neurogenesis. II. The role of cell interactions and cell lineage in the determination of neuronal precursor cells. Dev Biol 111:206–219.

Doe CQ Bastiani MJ Goodman CS (1986) Guidance of neuronal growth cones in the grasshopper embryo. IV. Temporal delay experiments. J Neurosci 6:3552–3563.

Doe CQ Hirómi Y Gehring WJ Goodman CS (1988) Expression and function of the segmentation gene *fushi tarazu* during *Drosphila* neurogenesis. Science 239:171–175.

Doe CQ Technau GM (1993) Identification and cell lineage of individual neural precursors in the *Drosophila* CNS. TINS 16:510–514.

Ebens AJ Garren H Cheyette B Zipursky SL (1993) The *Drosophila* anachronism locus: a glycoprotein secreted by glia inhibits neuroblast proliferation. Cell 74:15–27.

Edwards JS Swales LS Bate M (1993) The differentiation between neuroglia and the connective tissue sheath in insect ganglia revisited: the neural lamella and perineurial sheath cells are absent in a mesodermless mutant of *Drosophila*. J Comp Neurol 333:301–308.

Faeder IR Salpeter MM (1970) Glutamate uptake by a stimulated insect nerve muscle preparation. J Cell Biol 4:300–305.

Fallon JR (1985) Preferential outgrowth of central nervous system neurites on astrocyte and Schwann cells as compared with nonglial cells in vitro. J Cell Biol 100:198–207.

Fredieu JR Mahowald AP (1989) Glial interactions with neurons during *Drosophila* embryogenesis. Development 106:739–748.

Fredieu JR Mahowald AP (1992) Glycoconjugate expression during *Drosophila* embryogenesis ACTA. Anatomica 149:89–99.

Furst A Mahowald AP (1985) Differentiation of primary neuroblasts in purified neural cell cultures from *Drosophila*. Dev Biol 109:184–192.

Ghysen A Dambly-Chaudière C Aceves E Jan LY Jan YN (1986) Sensory neurons and peripheral pathways in *Drosophila* embryos. Roux's Arch Dev Biol 195:281–289.

Harrison RG (1910) The outgrowth of the nerve fiber as a mode of protoplasmic movement. J Exp Zool 9:787–846.

Hartenstein V Rudloff E Campos-Ortega JA (1987) The pattern of proliferation of the neuroblasts in the wild-type embryo of *Drosophila* melanogaster. Roux's Arch Dev Biol 196:473–485.

Hartenstein V Jan YN (1992) Studying *Drosophila* embryogenesis with P-lacZ enhancer trap lines. Roux's Arch Dev Biol 201:194–220.

Hoyle G (1986) Glial cells of an insect ganglion J Comp Neurology 246:85–103.

Huff R Furst A Mahowald AP (1989) *Drosophila* embryonic neuroblasts in culture: Autonomous differentiation of specific neurotransmitters. Dev Biol 134:146–157.

Jacobs JR Goodman CS (1989a) Embryonic development of axon pathways in the *Drosophila* CNS: I. a glial scaffold appears before the first growth cones J Neurosci 9:2402–2411.

Jacobs JR Goodman CS (1989b) Embryonic development of axon pathways in the *Drosophila* CNS. II. Behavior of pioneer growth cones. J Neurosci 9:2412–2422.

Jacobs JR Hiromi Y Patel NH Goodman CS (1989) Lineage, migration, and morphogenesis of longitudinal glia in the *Drosophila* CNS as revealed by a molecular lineage marker. Neuron 2:1625–1631.

Jan LY Jan YN (1982) Antibodies to horseradish peroxidase as specific neuronal markers in *Drosophila* and grasshopper embryos. PNAS 79:2700– 2704.

Klambt C Goodman CS (1991a) The diversity and pattern of glia during axon pathway formation in the *Drosophila* embryo. Glia 4:205–213.

Klambt C Goodman CS (1991b) Role of the midline glia and neurons in the formation of the axon commissures in the central nervous system of the *Drosophila* embryo. Ann NY Acad Sci 633:142–159.

Klambt C Jacobs JR Goodman CS (1991) The midline of the *Drosophila* central nervous system: a model for the genetic analysis of cell fate, cell migration, and growth cone guidance. Cell 64:801–815.

Klambt C Glazer L Shilo B-Z (1992) *breathless* a *Drosophila* FGF receptor homolog, is essential for migration of tracheal and specific midline glial cells. Genes Dev 6:1668–1678.

Letourneau PC (1975) Cell to subsratum adhesion and guidance of axonal elongation. Dev Biol 44:92–101.

Meyer MR Reddy GR Edwards JS (1987) Immunological probes reveal spatial and developmental diversity in insect glia. J Neurosci 7:512–521.

Nambu JR Franks RG Hu S Crews ST (1990) The single-minded gene of *Drosophila* is required for the expression of genes important for the development of the CN Smidline cells. Cell 63:63–75.

Nelson HB Laughton A (1993) *Drosophila* glial architecture and development: analysis using a collection of new cell-specific markers. Roux's Arch Dev Biol 202:341–354.

Noble M Murray K (1984) Purified astrocytes promote the in vitro divisions of a bipotential glial progenitor cell EMBO J 3:2243–2247.

Nusbaum J (1883) Vorläufige mittheilung über die chorda der Arthopoden. Zool Anzeig 6:291–295.

Patel NH Snow PM Goodman CS (1987) Characterization and cloning of fasciclin III: a glycoprotein expressed on a subset of neurons and axon pathways in *Drosophila*. Cell 48:975–988.

Poulson DF (1950) Histogenesis organogenesis and differentiation in the embryo of *Drosophila* melanogaster, in Biology of *Drosophila* (Demerec M, ed.) Wiley, New York, pp. 168–274.

Ramon y Cajal S Sanchez D (1915) Contribucion al conocimiento de los centros nerviosos de los insectos. Trabajos Lab Invest Biol Univ Madrid 13:1–164.

Remahl S Hildebrand C (1990) Relation between axons and oligodendroglial cells during initial myelination. I the glial unit. J Neurocytol 19:313–328.

Reynolds R Herschkowitz N (1986) Selective uptake of neuroactive amino acids by both oligodendrocytes and astrocytes in primary dissociated culture: a possible role for oligodendrocytes in neurotransmitter metabolism. Brain Res 371:253–266.

Rothberg JM Jacobs JR Goodman CS Artavanis-Tsakonas S (1990) slit: an extracellular protein necessary for development of midline glia and commissural axon pathways contains both EGF and LRR domains. Genes Dev 4:2169–2187.

Scharrer BCJ (1939) The differentiation between glia and connective tissue sheath in the cockroach *(Periplaneta Americana)*. J Comp Neurology 70:77–88.

Seeger M Tear G Ferres-Marco D Goodman CS (1993) Mutations affecting growth cone guidance in *Drosophila*: Genes necessary for guidance toward or away from the midline. Neuron 10:409–426.

Smith HK O'Kane CJ (1991) Use of a cytoplasmically localized P-lacZ fusion to identify cell shapes by enhancer trapping in *Drosophila*. Roux's Arch Dev Biol 200:306–311.

Smith PJS Shepard D Edwards JS (1991) Neural repair and glial proliferation: parallels with gliogenesis in insects. Bioessays 13:65–72.

Smith DS Treherne JE (1963) Functional aspects of the organization of the insect nervous system Adv Insect Physiol 1:401–478.

Smouse D Perrimon N (1990) Genetic dissection of a complex neurological mutant, polyhomeotic, in *Drosophila*. Dev Biol 139:169–85.

Sohal RS Sharma SP Couch EF (1972) Fine structure of the neural sheath, glia and neurons in the brain of the housefly, *Musca domestica*. Z Zellforsch Mikrosk Anat 135:449–59.

Springer CA Rutschky CW (1968) A comparative study of the embryological development of the median cord in Hemiptera. J Morph 129:375–400.

Thomas JB Bastiani MJ Bate M Goodman CS (1984) From grasshopper to *Drosophila*: a common plan for neuronal development. Nature 310203–207.

Thomas JB Crews ST Goodman CS (1988) Molecular genetics of the single-minded locus: a gene involved in the development of the *Drosophila* nervous system. Cell 52:133–141.

Treherne JE Smith PJS Edwards H (1987) Neural repair in an insect: Cell recruitment and deployment following selective glial disruption. Cell Tissue Res 247:121–128.

Truman JW Bate M (1988) Spatial and temporal patterns of neurogenesis in the central nervous system of *Drosophila* melanogaster. Dev Biol 125:145–157.

Tytell M Greenberg SG Lasek RJ (1986) Heat shock-like protein is transferred from glia to axon. Brain Res 363:161–164.

Udolf G Prokop A Bossing T Technau GM (1993) A common precursor for glia and neurons in the embryonic CNS of *Drosophila* gives rise to segment-specific lineage variants. Development 118:765–775.

Vanhems E (1985) An *in vitro* autoradiographic study of gliogenesis in the embryonic locust brain. Dev Brain Res 23:269–275.

Wheeler WM (1893) A contribution to insect embryology. J Morphol 8:2–160.

Yamada KM Spooner BS Wessells NK (1971) Ultrastructure and function of growth cones and axons of cultured nerve cell. J Cell Biol 496:14–35.

PART II

PHYLOGENY OF MYELINATION

The Evolution of Myelinating Cells

Betty I. Roots

1. Introduction

The development, maintenance, and functioning of the complex glial ensheathment of axons in vertebrates (phylum Chordata) is a classic manifestation of neuron–glia interrelations. Oligodendrocytes form the sheath in the central nervous system (CNS) and Schwann cells in the peripheral nervous system (PNS)—cells that differ in their origin and in their nature. Complex glial ensheathment of axons is not restricted to vertebrates, however, but is found also in a number of invertebrate species in the phyla Annelida and Arthropoda. In all the organisms in which they are found, myelin and myelin-like sheaths enhance the conduction velocity of action potentials along the axons by making saltatory conduction possible. In all sheaths, vertebrate and invertebrate alike, there are interruptions that mediate the jumping of the potential along the nerve fiber. These interruptions may be simply the point of origin of branches or well-developed nodes, such as the nodes of Ranvier in vertebrates. In this chapter the occurrence of myelin and myelin-like sheaths in the three phyla will be surveyed, the sheaths and the myelinating cells that give rise to them will be compared, and their evolutionary significance will be discussed.

2. Occurrence of Myelin and Myelin-Like Sheaths

2.1. Chordata

Myelinated axons are found in both the peripheral and central nervous systems of fishes, amphibia, reptiles, birds and mammals, i.e., in all of the vertebrates except lampreys and hagfishes

From: *Neuron-Glia Interrelations During Phylogeny: I. Phylogeny and Ontogeny of Glial Cells*
A. Vernadakis and B. Roots, Eds. Humana Press Inc., Totowa, NJ

(Agnatha) (Bullock et al., 1984). In other groups of chordates myelin sheaths are also lacking. (*See* Fig. 1 for the phylogenetic relationships of animals.) In the peripheral nerves of mammals, in general axons 1 μm or more in diameter are clothed in a multilayered sheath formed by processes of Schwann cells (Fraher, 1972; Hahn et al, 1987). In the CNS axons smaller than 1 μm may be myelinated and there is a greater overlap in diameters of myelinated and unmyelinated axons (Fraher, 1976; Remahl and Hildebrand, 1982). These sheaths are made by oligodendrocytes.

As has been well documented, compact vertebrate myelin in both PNS and CNS consists of multiple spirally wrapped lamellae of glial processes in which both the extracellular space between lamellae and the glial cytoplasm are virtually excluded so that the cell membranes are closely apposed. In most, but not all, myelinated fibers the sheath is interrupted at intervals leaving an unmyelinated portion of axolemma that extends for about 1–1.5 μm in the PNS, and 1–10 μm or more in the CNS. At these nodes various membrane specializations are found. The morphology of myelin and nodes of Ranvier is remarkably constant in all vertebrate classes, but in the fishes and amphibia myelination is less extensive than in mammals, birds, and reptiles.

It should be noted that neuronal somata may also be myelinated, notably in the vestibular and spiral (acoustic) ganglia in mammals (Rosenbluth, 1962; Ona, 1993) and goldfish (Rosenbluth and Palay, 1961), and in chick ciliary ganglia (Hess, 1965). Both compact myelin and "loose" myelin, in which varying amounts of glial cytoplasm together with inclusions, such as mitochondria, are retained in the glial processes, are found around neuronal somata. Myelinated dendrites are found in the olfactory bulb of mammals (Pinching, 1971; Willey, 1973).

2.2. Annelida

Multilayered myelin-like sheaths are found around the giant fibers in two classes of annelids, the polychaeta and the oligochaeta. Well-developed sheaths are found in a number of species belonging to three families of polychaetes. *Mastobranchus* sp (Capitellidae), *Prionospio steenstrupi* (Spionidae), and *Clymene producta* and *C. torquata* (Maldanidae) (Nicol, 1948). The ultrastructure of these sheaths has not been described.

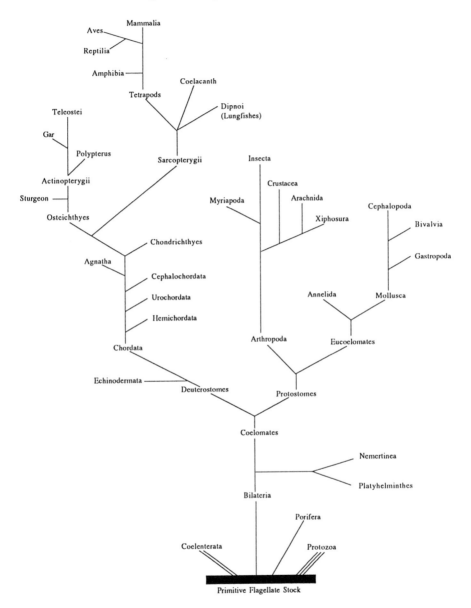

Fig. 1. Phylogenetic tree (from Roots, 1993).

Spirally wrapped sheaths bearing a remarkable resemblance to vertebrate myelin have been described in three families of oligochaeta, the Lumbricidae, Tubificidae, and Lumbriculidae. In earthworms (Lumbricidae) the sheaths surrounding the median and two lateral giant fibers have been described in two species, *Eisenia foetida* (Hama, 1959) and *Lumbricus terrestris* (Coggeshall, 1965; Roots and Lane, 1983). The spacing between the glial membranes is irregular and, whereas in some regions glial cytoplasm is almost completely extruded, in others, especially where the sheath is buckled, considerably more cytoplasm is present. The median giant fiber is more heavily myelinated than the laterals and in *Lumbricus terrestris* has as many as 200 lamellae.

A striking characteristic of Lumbricid sheaths is the bands of desmosome-like structures running serially across the lamellae. (*see* Fig. 2). In *Lumbricus terrestris*, there are two circular pores in the sheath in each segment of the animal (Günther, 1973, 1976; Roots, 1984) that functionally are the equivalent of nodes of Ranvier in vertebrate sheaths (Günther, 1976). The paired lateral giant fibers of the tubificid worm, *Branchiura sowerbyi*, have similar sheaths with redundant loops and swirls and desmosomal stacks, but no dorsal nodes were described (Zoran et al., 1988). In their paper on escape reflexes of an aquatic oligochaete, *Lumbriculus variegatus* (Lumbriculidae), Drewes and Brinkhurst (1990) include an electron micrograph of giant fibers of a newly hatched worm showing a "developing myelin-like sheath." No description is given but the sheath appears to be similar to that found in other oligochaetes.

2.3. Arthropoda

In the arthropoda, myelin-like sheaths are found only in the decapod crustacea (shrimps, prawns, lobsters, crayfish, and crabs) where many nerve fibers, especially those of giant motor axons, are invested with a sheath that by light microscopy bears a striking resemblance to the myelin sheaths of vertebrates, including interruption by nodes (Retzius, 1890; Nageotte, 1916; Johnson, 1924; Bear and Schmitt, 1937; Holmes, 1942). Examination by electron microscopy, however, shows that the sheath differs radically from that of both vertebrates and annelids. The sheath is not spirally wound but consists of concentric lamellae. The description is most complete for a prawn, *Palaemonetes vulgaris* (Heuser and Doggenweiler, 1966). In this species the laminae vary between 10 and 200 nm in thickness, the thinner ones being toward the outer surface of the sheath. Where the ends of each lamina meet, a short seam, or mesaxon, is formed. These seams are arranged in a regu-

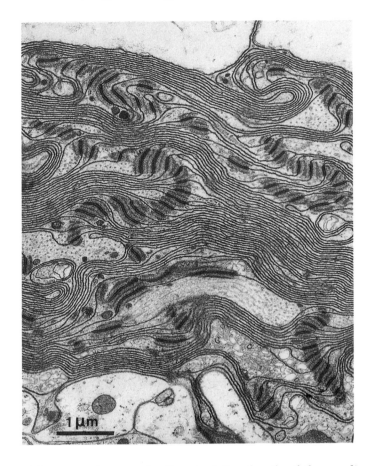

Fig. 2. Electron micrograph of part of the sheath of the median giant fiber of the earthworm, *Lumbricus terrestris*. Note the desmosome-like structures running in register across the sheath, and pockets of cytoplasm.

lar pattern, with those of alternate laminae being on opposite sides of the axon. Cytoplasm is generally found within the glial cell processes (*see* Fig. 3), although in the outermost layers of the thickest sheaths both it and the extracellular space between the processes may be obliterated, producing close membrane appositions analogous to the major and intraperiod lines, respectively, of vertebrate myelin. Desmosome-like structures occur in register through the thickness of the sheath, reminiscent of the structure of the annelid sheath. The sheath is interrupted by annular nodes that, however, are different in structure from those found in vertebrates.

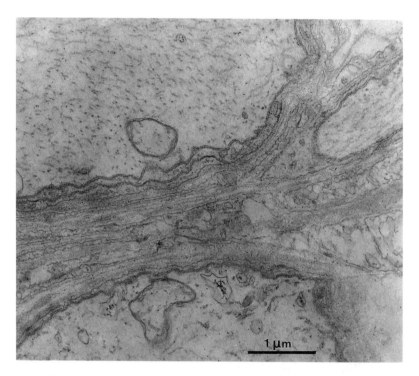

Fig. 3. Electron micrograph of the loose-myelin sheaths of two neighboring axons in the ventral nerve cord of the crayfish, *Procambarus clarki*.

(For a review and discussion of the evolution of nodes, *see* Roots,
1984.) Other major differences between the sheath in prawns and
that in vertebrates is the presence of a glial cell layer with nuclei
between the axolemma and the sheath, and the absence of glial
cell nuclei between the outside of the sheath and the surrounding
connective tissue.

In the CNS of the crab, *Cancer irroratus*, sheaths bearing a
very close resemblance to compact vertebrate myelin with glial
cell nuclei outside the sheath, structures resembling Schmidt-
Lantermann incisures and with nodes strikingly like vertebrate
nodes of Ranvier have been described (McAlear et al., 1958).

The arrangement is rather different in the shrimp, *Penaeus
japonicus*, in that the myelin sheath does not wrap the axon directly
but encloses an extracellular space in which the axon with its
associated glial cell, lies attached to the inner surface of the myelin
sheath by a connective. The glial cell associated with the axon is

packed with microtubules (Hama, 1966). The myelin sheath was not described but from the figures appears to be quite thick, about 17 μm with the axon being <1 μm.

3. Myelinating Cells in Chordata

3.1. Schwann Cells

3.1.1. Mammals

During development, as the neural plate rolls up to form the neural tube, cells on either side in a dorsolateral position do not become part of the tube but form the neural crest. It is from the cells of these two longitudinal bands that Schwann cells originate (*see* Hörstadius, 1950, Le Douarin, 1982, for reviews). As the future Schwann cells migrate out from the crest, they become spindle shaped. They later develop short thick processes from near the main body of the cell that come to segregate groups of tightly packed axons. As migration ends, proliferation of the cells occurs. Developing Schwann cells have a large volume of cytoplasm rich in mitochondria, polyribosomes, and rough endoplasmic reticulum. Bundles of microtubules parallel to the long axes of the axons with which they are associated are present, as are some microfilaments. Mature Schwann cells have elongated nuclei and the organelles are concentrated in, but not restricted to, the perinuclear cytoplasm. There is a well-developed Golgi apparatus, filaments are more in evidence but there is less rough endoplasmic reticulum. Lysosomes, glycogen, lipid droplets, and multivesicular bodies are also present. Schwann cells have the potential to develop into either a myelinating cell or an ensheathing cell determined by the signals received from the axons.

Myelinating Schwann cells achieve a 1:1 ratio with axons, cease to proliferate and begin myelination covering the axon with spirally wrapped overlapping layers of cell processes. In mature myelinated fibers, the Schwann cell nucleus is located at the midpoint of the cell and only one or more thin strips of Schwann cell cytoplasm remain. Additional cytoplasm is found in the Schmidt-Lantermann clefts that traverse the sheath at intervals providing a helical connecting pathway between inner and outer layers of Schwann cell cytoplasm. At the nodes of Ranvier, finger-like processes project into the nodal gap substance.

Nonmyelinating Schwann cells ensheath axons in deep troughs, each Schwann cell encompassing a varying number of axons. As in myelinating Schwann cells, the nucleus is situated at the midpoint of the cell, but the cells are about one-tenth the length of myelinating Schwann cells (Schnapp and Mugnaini, 1975; Landon and Hall, 1976; Ochoa, 1976; Raine, 1984; Jessen and Mirsky, 1991; Peters et al., 1991).

Antigenically mature myelinating and nonmyelinating Schwann cells present different phenotypes. The antigenic profile of precursor cells has not yet been completely determined, but the developmental sequence from before the cells are committed to the myelinating or nonmyelinating phenotype to maturity has been established for rat limb nerves. A brief outline is given here. (For a full discussion the reader is referred to Jessen and Mirsky, 1991; Gould et al., 1992; Jessen et al., 1994, and Chapter 8 by Stewart and Jessen.)

At the time of the transition from precursor cells to Schwann cells (day E15-17), S-100 β is expressed. Shortly after the first appearance of S-100, the O4 antigen that marks sulfatide appears. The differentiation into myelinating and nonmyelinating Schwann cells begins at day E18 when the future myelinating cells begin to express galactocerebroside (Gal-C).

Further differentiation occurs as myelination begins around birth. The expression of molecules such as the NGF receptor and the surface proteins L1, NCAM, A5E3, and Ran-2, are downregulated in the future myelinating Schwann cells (they persist in the nonmyelinating cells) and at the same time the myelin proteins MAG, P_0, and MBP are expressed. Interestingly, in the third postnatal week Gal-C is expressed by nonmyelinating Schwann cells.

In order to myelinate an axon, Schwann cells must form a basal lamina. Fernandez-Valle et al. (1993) showed that basal lamina synthesis induces both changes in gene expression and cell shape and restricts expression of myelin genes to the Schwann cells associated with myelin-requiring axons.

A detailed account of the composition of the multiple membranes of myelinating Schwann cells, myelin, will not be given here but certain salient points will be mentioned. (For reviews of the composition of myelin, both peripheral and central, across animal phyla, see Bürgisser et al., 1986; Waehneldt, 1990; Kirschner and Blaurock, 1992; Roots, 1993.)

The major lipids of mammalian peripheral myelin are cholesterol, phospholipids, and the galactosphingolipids, cerebroside and sulfatide. The principal protein is P_0, a glycoprotein with a mol wt of 30 kDa, which comprises more than half the total myelin protein and which has a large hydrophobic region. Myelin basic proteins (MBPs) P_1, P_r, and P_2 are present in varying amounts in various mammalian species. A high molecular weight MAG is present, as is a glycoprotein of 170 kDa that is specific for peripheral myelin. CNPase is present but in lower amounts than in central myelin.

3.1.2. Other Vertebrates

The origin of Schwann cells from the neural crest is well documented for birds and amphibia, indeed much of the experimental work has been carried out using species from these groups (Hörstadius, 1950; Le Douarin, 1982; Le Douarin et al., 1991). It is probably safe to assume that since there are Schwann cells in reptiles and fishes and a neural crest during development, their origin in these groups is also from the neural crest, but the details of their development may well be different, especially in the fishes.

Studies on the antigenic phenotype of Schwann cells in groups other than mammals are only just beginning. In birds two antibodies, one against the 4B3 epitope, a carbohydrate moiety associated with a variety of molecules, and another against Schwann cell myelin protein (SMP) which appears as a doublet of 75 and 80 kDa on SDS-PAGE gels, provide early markers of peripheral glial cells. The anti-SMP appears to be specific for Schwann cells, both myelinating and nonmyelinating (Le Douarin et al., 1991).

There is evidence that Schwann cells in goldfish express S-100 (Nona et al., 1992; *see also* Chapter 14 in *Neuron-Glia Interrelations During Phylogeny, Part II*).

The lipid composition of peripheral myelin is remarkably constant in overall pattern throughout the vertebrates, but there are significant differences in the proteins. In the fishes P_0 is replaced by several IPs (intermediate proteins) that, like P_0, are glycosylated and bind P_0 antibodies. Equivalents of the MBP P_1 are found in all groups but P_2 is found only in reptiles and birds apart from mammals. The presence of MAG has not been established unequivocably for fishes. CNPase is very low in fishes.

3.2. Oligodendrocytes

3.2.1. Mammals

Cajal first showed that neuroglia were derived from the lining layer of the brain central canal (Cajal, 1901–1917). That oligodendrocytes originate from neuroectodermal cells of the subventricular zone in the forebrain and cerebellum and that the progeny of these cells migrate into both grey and white matter undergoing further proliferation and differentiation has been firmly established (Privat, 1975). What happens in their further development and in the spinal cord, however, is by no means so clearly established.

It was suggested by Choi and his co-investigators (Choi et al., 1983; Choi and Kim, 1985) that in the spinal cord oligodendrocytes arise from embryonic radial glial cells. This hypothesis is supported by Aloisi et al., 1992, but other recent evidence does not support it (Warf et al., 1991; Noll and Miller, 1993; Pringle and Richardson, 1993), so the issue cannot be considered to have been settled.

Raff et al. (1983) found that in cultures of dissociated cells prepared from developing rat optic nerves there were cells that had the potential to develop into either oligodendrocytes or a type of astrocyte depending on culture conditions. He termed these cells 0-2A progenitor cells. One question that has been asked is whether 0-2A progenitor cells are found in vivo. Certainly oligodendrocyte progenitor cells have been identified in vivo (see, for example, Reynolds and Wilkin, 1988, 1991; Hardy and Reynolds, 1991) and have been isolated from the brain (Dubois-Dalcq, 1987; Lubetzki et al., 1991). However, there is no evidence that these cells are bipotential in vivo and indeed it appears that 0-2A progenitors from culture develop only into oligodendrocytes but not astrocytes when transplanted into neonatal rat brains (Espinosa de los Monteros et al., 1993). For a review of this interesting question, *see* Williams and Price, 1992.

Morphologically, the early progenitor cells have a simple bipolar shape. As they differentiate, they develop more processes. In mature oligodendrocytes, the nucleus is usually round or oval with clumping of the chromatin, particularly at the periphery, and is eccentrically placed in the cell. The cytoplasm is rich in organelles. The agranular endoplasmic reticulum is prominent, rough endoplasmic reticulum well developed, and microtubules are a characteristic component. Mitochondria, free ribosomes, multivesicular bodies, and lamellar dense bodies are distributed throughout the cytoplasm. Filaments are absent (Mugnaini and Walberg, 1964; Cammermeyer, 1966). Mori and Leblond (1970) described three types of oligodendrocytes, light, medium, and dark, depending on the staining properties of their cytoplasm. They suggested that these different forms represent stages in oligodendrocyte differentiation. The light cells take up tritiated thymidine and are actively dividing. As the oligodendrocytes mature, they become progressively darker. This proposed progression has been confirmed by later studies (*see* Skoff, 1980; Parnavelas et al., 1983) and by recent studies using the antibody 04 *(see later)* to identify mature oligodendroglia (Warrington and Pfeiffer, 1992). The mature, dark cells are the most common (40%) in the adult brain.

Unlike Schwann cells, which myelinate only one internode, oligodendrocytes myelinate internodes of many axons, as many as 60 in rat spinal cord (Matthews and Duncan, 1971), and may be separated by long, slender processes from the axon they are myelinating in contrast to Schwann cells that are in close apposition to the axon. As in the PNS the glial lamellae are spirally wrapped. Unlike Schwann cells, oligodendrocytes do not form a complete collar of cytoplasm around the outer surface of the sheath, but tongue-like processes cover only part of the fiber (Bunge, 1968). Schmidt-Lantermann clefts are rare and oligodendrocyte processes are not present at the nodes of Ranvier.

The antigenic profile of the cells changes during oligodendrocyte development. Early progenitor cells (0-2A cells) are positive to the monoclonal antibody A2B5 (which recognizes epitopes on several gangliosides) and they express the ganglioside GD_3, the NG-2 chondroitin sulfate and vimentin. As they develop more processes, the cells express sulfated glycolipids, recognized by the monoclonal antibody 04, A007 (a surface antigen), and CNPase. Later, as they approach maturity, they begin to express Gal-C, MBP, and the proteolipid myelin protein (PLP) and lose their antigenicity to A2B5 and GD3. Mature oligodendrocytes retain their 04 positive antigenicity. The precise timing of the antigenic sequence depends on environmental factors and is different in vivo and in culture. What is clear is that vimentin is expressed in the early stages of differentiation and that CNPase is also an early marker.

The synthesis of the myelin proteins MBP and PLP occurs at a relatively late stage of development. (*See* Behar et al., 1988; Hardy and Reynolds, 1991; and the review by Goldman, 1992.)

Like Schwann cells, oligodendrocytes require signals from axons for myelination but there are reports that they can produce myelin whorls or even sheaths (Althaus et al., 1987) in culture in the absence of axons. These studies indicate that oligodendrocytes are less dependent on axons to produce myelin than Schwann cells (Lubetzki et al., 1993). This conclusion is supported by the finding that oligodendrocytes in culture in the absence of axons express the major constituents of myelin and on a similar time course (Dubois-Dalcq et al., 1986), whereas Schwann cells do not have this capability (Brockes et al., 1979; Mirsky et al., 1980).

The lipid composition of CNS myelin is similar to that of peripheral myelin, differing only in the relative proportions of some components, notably the galactosphingolipid. There is, however, a major difference in the proteins. The major component is PLP, a protein of 30 kDa, which replaces P_0 of peripheral myelin. It is very hydrophobic but is not glycosylated. Another protein specific to CNS myelin is DM-20 (kDa 20), usually seen as a doublet on SDS-PAGE gels. As in PNS myelin, there are a number of MBPs depending on species. There is more CNPase than in PNS myelin and MAG is present.

3.2.2. Other Vertebrates

Oligodendrocytes remarkably similar in structure to those in mammals are found in birds (Lyser, 1972), reptiles (Kruger and Maxwell, 1967; Monzon-Mayor et al., 1990), amphibia (Stensaas and Stensaas, 1968a,b; Schonbach, 1969; Stensaas, 1977), and teleosts (Kruger and Maxwell, 1967; Jeserich and Waehneldt, 1986; Wolburg and Bouzehouane, 1986; Bastmeyer et al., 1989, 1991). (*See also* Chapters 4, 5, and 6 in this volume and Chapters 14 and 16 in *Neuron-Glia Interrelations During Phylogeny, Part II.*)

In the lizard, *Gallotia galloti*, light, medium, and dark oligodendrocytes together with cells termed active oligodendrocytes most numerous from E37 to hatching, the period of rapid myelination, have been described. The active oligodendrocytes have more organelles, especially more rough endoplasmic reticulum, than the lizard medium oligodendrocytes and approximate more closely to the medium oligodendrocytes described by Mori and Leblond (1970) (Monzon-Mayor et al., 1990). There are thus variations in oligodendrocyte structure but the most apparent difference is the relative paucity of oligodendrocytes in amphibia and fishes.

The composition of the myelin produced by oligodendrocytes in birds and reptiles is basically the same as that in mammals, but in the other classes there are important differences in the proteins. In amphibia DM-20 is not present but in both cartilaginous and teleost fishes neither DM-20 nor PLP are present. Instead, there are a varying number of glycosylated proteins referred to as intermediate proteins (IP) that furthermore bind antibodies to mammalian P_0. Thus, in fishes both central and peripheral myelins contain glycosylated proteins. Teleosts are further distinguished by having a 36-kDa nonglycosylated hydrophobic protein in central myelin. CNPase is very low in fish myelin.

The central myelin of the lungfishes, which are considered to be closer to the tetrapod ancestry than other fishes, contains a 29-kDa protein that reacts with PLP antibodies but is glycosylated. The coelacanth, *Latimeria*, which is even closer to the tetrapod ancestry, has a major component that comigrates with PLP and binds PLP antibodies. As in other fishes, DM-20 is lacking and CNPase is extremely low.

The only lower vertebrate group in which studies of lineage and developmental profile have been made is the teleosts. The findings are discussed in detail by Jeserich et al. in Chapter 11; therefore, only a brief outline will be given here. In trout IP2 and Gal-C are expressed by oligodendrocytes at stage 28 (8 d before hatching), followed first by 36 kDa, then IP1 (Jeserich et al., 1990). MBP could not be detected in the oligodendrocytes, but in tissue sections it appears in parallel with IP2. This sequence is the same as in myelin. Initially, developing oligodendrocytes are positive to both A2B5 and 6D2 (which recognizes the oligosaccharide moiety of IP1 and IP2 proteins), but as they mature, they lose their A2B5 antigenicity. In culture trout oligodendrocytes rapidly lose the ability to express myelin proteins, so in this respect they resemble mammalian Schwann cells (Jeserich and Rauen, 1990; Jeserich and Stratmann, 1992; Jeserich and Waehneldt-Kreysing, 1992; Sivron et al., 1992).

4. Myelinating Cells in Annelida and Arthropoda

4.1. Annelida

Little is known about the origin of glial cells in annelids. In the earthworms, the ventral nerve cord arises from two rows of ectodermal cells. Swellings that become the ganglia develop consecutively in all segments beginning at the anterior end. Concrescence of the two halves to form the ventral nerve cord also begins in the first segment. Sometime after the two halves have grown together, the giant fibers on the dorsal surface of the cord arise by differentiation from a fibrillar mass. Other than that they originate in this ectodermal structure, nothing is known of the development of glial cells (Wilson, 1889).

The glial cells that form the myelin-like sheath round the giant fibers of earthworms have relatively clear cytoplasm containing mitochondria, granular endoplasmic reticulum, a small Golgi apparatus, vesicles, and granules of various sizes, occasional tubules and bundles of fine 5-nm diameter filaments. The nuclei are usually located in the surrounding supportive glial area but may be found in one of the outer layers of the sheath (Hama, 1959; Coggeshall, 1965; Roots and Lane, 1983). By transmission electron microscopy, these cells cannot be distinguished from surrounding supportive glial cells (Coggeshall, 1965), but freeze-fracture studies have shown that the density of intramembranous particles is about $80/\mu m^2$ in the myelinating glial cells and much higher, about $400/\mu m^2$, in the nonmyelinating cells. Furthermore, the size of the particles is smaller in the myelinating glia, 8.13 nm in diameter, compared with 9.76 nm (Roots and Lane, 1983).

No information is available regarding the lipid composition of the glial myelinating membranes, but from analyses of the whole nerve cord it may be deduced that they contain cholesterol and phospholipids but no galactocerebroside or sulfatide. Traces of glucocerebroside are found, however. Interestingly, no sphingomyelin, which has been regarded as ubiquitous in biological membranes, was detected (Okamura et al., 1985 a, b).

The protein composition of the membranes forming the myelin-like sheath is completely different from that of any vertebrate myelin. The profile is relatively simple, with the major components being 80- and 42-kDa proteins and minor proteins of 28–32 kDa. No crossreactivity was found with antibodies generated against mammalian MBP, PLP, MAG, or CNP. In fractions identified as originating from the desmosome-like structures, proteins in the 24–34 range were enriched rather than the 85–150 kDa, which would be expected from desmosomes (Pereyra and Roots, 1988; Cardone and Roots, 1990). This supports the morphological observations that the structures are not true desmosomes (Roots and Lane, 1983).

Antibodies were raised to the membrane fraction. A polyclonal antibody reacted predominantly with the 40–42-kDa proteins and a monoclonal antibody bound to the 42-kDa and to a 30-kDa protein. In immunogold electron microscopical studies the antibodies bound to both the myelin-like membranes and to membranes of other glial processes (*see* Fig. 4). Thus, these proteins are common to both myelinating and nonmyelinating glia (Cardone and Roots, 1991).

4.2. Arthropoda

As in the annelids, little is known about the origin of glial cells other than that it is ectodermal.

The cytoplasm in the glial lamellae constituting the sheath in the prawn, *Palaemonetes vulgaris*, contains mitochondria and many longitudinally oriented microtubules. The nucleus of the cell that forms the sheath is located in the innermost lamina (Heuser and Doggenweiler, 1966). The authors discuss the possible nature of the connections between the concentric laminae and conclude that only developmental studies would be able to determine their nature and formation. Such studies have not as yet been carried out.

The glial cells ensheathing the giant fibers of the spiny lobster, *Panuliris argus*, contain microtubules and actin-like filaments are prominent (Warren and Rubin, 1987). The glial lamellae wrapping axons of crayfish (*Procambarus* and *Cambarus* species) contain pale cytoplasm containing few microtubules, mitochondria, Golgi, and rough and smooth endoplasmic reticulum (Norlander and Singer, 1972; and the author's observations). A feature of both lobster (Holtzman et al., 1970) and crayfish (Peracchia and Robertson, 1971; Lane and Abbott, 1975; Shivers and Brightman, 1976; Cuadras and Marti-Subirana, 1987) ensheathing glia cells is the tubular lattice system that transverses them. The channels are approx 25 nm wide and 50 nm long with a density of approx

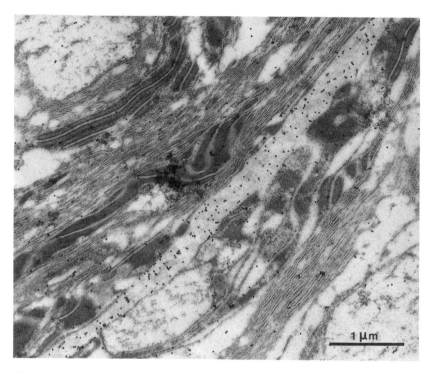

Fig. 4. Immunogold staining of earthworm giant fiber showing gold particles bound to myelin membranes and associated non-myelinating glia.

$16/\mu m^2$. The channels are not precisely aligned between glial layers but they do provide a channel that transverses the loose-myelin sheath.

The lipid composition of the glial membranes ensheathing the axons of the pink shrimp, *Penaeus duorarum*, has been determined. They are rich in lipids with a 15:1 lipid:protein ratio with cholesterol and phospholipids being the major constituents. No galactocerebroside was found but there are substantial amounts of glucocerebroside, about the same proportion as galactocerebrosides in mammalian myelin. Sphingomyelin is present, but its structure was found to be different from that in mammals (Okamura et al., 1986).

The protein profile has also been determined for the myelin-like membranes of *Penaeus duorarum*. There are four major proteins, molecular weights 21.5, 40, 78, and 85 kDa, and four minor proteins, mol wt 36, 41.5, 43, and 50 kDa. None of the proteins showed cross-reactivity with mammalian MBP or PLP, or with trout BP_1, 36 kDa or IP1 antibodies (Okamura et al., 1986; Waehneldt et al., 1989). Weak crossreactivity to bovine P_0 antisera was shown by the 50-kDa protein. However, it has not been established conclusively that the 50-kDa protein is a constituent of the myelin-like membranes. Moreover, a 50-kDa protein, which also shows P_0 crossreactivity, has been found in the CNS of hagfish and lamprey in which there is no myelin sheath (Waehneldt et al., 1989; Waehneldt, 1990).

We tested the antibodies we raised against earthworm myelin membrane proteins against proteins in crayfish (*Procambarus clarki*) nerve cord homogenate and found crossreactivity to 60–65-, 42-, and 40-kDa proteins and to a lesser extent to a 30-kDa protein (Cardone and Roots, 1991). In immunogold electron microscopical studies, the antibodies bound to the sheath membranes but not the axonal membrane (*see* Fig. 5). Thus, some antigenic properties are shared by the sheaths of earthworm and crayfish but there is no apparent resemblance to vertebrate myelin proteins.

5. Conclusions

Naked axons are common only in the coelenterates, in all other phyla axons have some kind of glial sheath. This may be a simple investment of glial cytoplasm, but overlapping glial lamellae are common in the invertebrate phyla and lower chordates and are found surrounding not only axons but neuronal somata as well (Roots, 1978). It is thus easy to imagine how the more elaborate loose-myelin sheaths of crustacea and annelids, and the more complex compact myelin sheaths of vertebrates, may have evolved. It is nevertheless a most remarkable example of convergent evolution that myelinating glial cells have evolved in three such disparate phyla.

The advantages of myelin sheaths on axons, particularly ones with nodes, are clear. Not only is the conduction velocity of action potentials increased but it is achieved with economy of space vis-à-vis increasing axon diameter and with saving of energy since electrotonic events occur only at interruptions in the sheath, at branching or at nodes. Conduction velocity in the giant unmyeli-

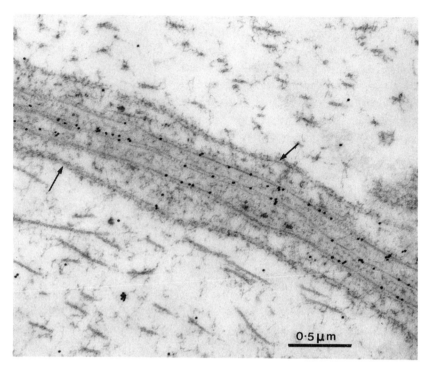

Fig. 5. Immunogold staining of crayfish nerve cord showing gold particles bound to the glial membranes ensheathing the axons but not to the axolemmas (arrows).

nated fibers of squid 500 μm in diameter is 20 m/s (from Prosser, 1961), whereas in the median giant fiber of the earthworm of only 90 μm in diameter it is 30 m/s (Günther, 1976), which may be attributed to the myelin-like sheath on the earthworm fiber. In the shrimp, *Penaeus japonicus*, a velocity of 90–219 m/s is achieved in fibers 120 μm in diameter. The difference between earthworm and shrimp may be attributable to the functional efficiency of the nodes. Günther (1976) found that the nodes in earthworm reduced the conduction time over a distance of 20 cm by only 3.5 ms. The compact sheaths of vertebrates are far more effective in increasing conduction velocity, for example, in a rat fiber 4.5 μm in diameter and with a sheath about 50% of its total diameter the conduction velocity is 59 m/s.

Evolutionary trends are most apparent in the biochemical characteristics of myelinating cells. As we have seen, galactosphingolipids, cerebroside and sulfatide, are major components of the lipids of both peripheral and central myelin in all the vertebrate classes, whereas in shrimp glucocerebroside rather than galactocerebroside is present. In earthworms, likewise, there are no galactosphingolipids but traces of glucocerebroside are present in the nerve cord. Okamura et al. (1985b) examined the lipids in the nervous systems of a wide variety of species across the spectrum of animal phyla. They found that only glucocerebrosides are found in protostomes, whereas in the deuterostomes galactocerebrosides and galactosulfatide are predominant. Deuterostomes retain the ability to synthesize glucocerebroside as is evidenced by the occurrence of considerable amounts of glucocerebroside in gadoid fishes (Tamai et al., 1992). There was, however, an early evolutionary divergence with the deuterostomes acquiring the ability to synthesize galactosphingolipids.

The proteins synthesized by myelinating glial cells in annelids and crustacea are very different from those in vertebrates. Although testing has not been exhaustive and the proteins have not been characterized, they appear to have nothing in common with any of the vertebrate myelin proteins. Earthworm and crayfish proteins have some antigenic epitopes in common, but since nothing is known of the amino acid sequence and structure of the proteins, the significance of this fact remains obscure.

In the vertebrates the major evolutionary trend has been the increasing differentiation of oligodendrocytes in the CNS with the evolution of different protein constituents in the membranes of the myelinating cells. In the fishes, cartilaginous and bony, both Schwann cells and oligodendrocytes elaborate membranes containing glycosylated hydrophobic proteins, IPs. In amphibia, oligodendrocytes synthesize PLP, a nonglycosylated hydrophobic protein instead. This represents a major evolutionary step. DM-20, an alternatively spliced variant of the PLP gene product, appears as a later development in reptiles, birds, and mammals. The 36-kDa hydrophobic protein of teleosts seems to be an evolutionary branch. Very little CNPase is synthesized by fish Schwann cells and oligodendrocytes but those of amphibia, reptiles, birds, and mammals produce it in larger amounts, oligodendrocytes making more than Schwann cells.

It is interesting to consider how apparent sudden leaps in protein synthesis, such as the production of PLP by amphibian oligodendrocytes may occur in evolution. There is evidence that the PLP gene extends back in evolutionary history to the chondrichthyes in which group vertebrate myelin first appears. Three proteins have been identified in *Torpedo* and the spiny dogfish, *Squalus acanthias*, which are closely related to each other and to mammalian DM-20 in nucleotide sequence. Two similar proteins were found in the zebrafish, *Branchiodanio rerio*, a teleost. The newly described DM proteins are the products of a gene that from sequencing data has been shown to be related to the PLP gene. It was suggested that one of the proteins DM_α is ancestral to PLP and DM-20. This DM_α protein has substantial sequence similarity with some proteins that form pore complexes across neuronal membranes, for example, the nicotinic acetylcholine receptor in *Torpedo* (Kitagawa et al., 1993). This may well be an example of a gene that, during the course of evolution, is modified with its products adapted to new functions. Supporting this view are several facts. One is that DM-20 is expressed early in ontogeny, before PLP, and appears to play a role in oligodendrocyte differentiation, possibly by involvement in a signaling pathway (Hudson and Nadon, 1992; Tamsit et al., 1992). This role appears to precede any other role in myelination in evolution. Another is that PLP has a dual function, being involved in oligodendrocyte maturation as well as myelin compaction and also possesses the properties of an ionophore. Thus, it is possible that genes (and their promoters) responsible for transmembrane signalling may, during the course of evolution, have been modified and rearranged for different functions.

Further studies on gene expression in myelinating cells throughout the vertebrates and the application of molecular genetics to the study of invertebrate myelinating cells should be very rewarding and raise more interesting questions.

Acknowledgments

The work done in the author's laboratory was supported by Grant No. A6052 from the Natural Sciences and Engineering Research Council of Canada. The author thanks Beatrice Cardone and Richard Cameron for technical assistance and Ed Knapp for photographic assistance, and is grateful to JAI Press Inc. for permission to reproduce Fig. 1.

References

Aloisi F Giampaolo A Russo G Peschle C Levi G (1992) Developmental appearance, antigenic profile, and proliferation of glial cells of the human embryonic spinal cord: An immunocytochemical study using dissociated cultured cells. Glia 5:171-181.

Althaus HH Bürgisser P Kloppner S Rohmann A Schroter J Schwartz P Siepl C Neuhoff V (1987) Oligodendrocytes ensheath carbon fibres and produce myelin *in vitro*. NATO ASI Series H, vol. 2. Glial-Neuronal Communication in Development and Regeneration, Springer-Verlag, New York, pp. 781-798.

Bastmeyer M Beckmann M Nona SM Cronly-Dillon JR Stuermer CAO (1989) Identification of astrocyte- and oligodendrocyte-like cells of goldfish optic nerves in culture. Neurosci Lett 101:127-132.

Bastmeyer M Beckmann M Schwab ME Stuermer CA (1991) Growth of regenerating goldfish axons is inhibited by rat oligodendrocytes and CNS myelin but not by goldfish optic nerve tract oligodendrocyte-like cells and fish myelin. J Neurosci 11:626–640.

Bear RS Schmitt FO (1937) Optical properties of the axon sheaths of crustacea nerves. J Cell Comp Physiol 9:275–287.

Behar T McMorris FA Novotny EA Barker JL Dubois-Dalcq M (1988) Growth and differentiation properties of 0-2A progenitors purified from rat cerebral hemispheres. J Neurosci Res 21:168–180.

Brockes JP Raff MC Nishiguichi DH Winter J (1979) Studies on cultured Schwann cells. III. Assays for peripheral myelin proteins. J Neurocytol 9:66–77.

Bullock TH Moore JK Fields D (1984) Evolution of myelin sheaths: Both lamprey and hagfish lack myelin. Neurosci Lett 48:145–148.

Bunge RP (1968) Glial cells and the central myelin sheath. Physiol Rev 48: 197–251.

Bürgisser P Matthieu J-M Jeserich G Waehneldt TV (1986) Myelin lipids: a phylogenetic study. Neurochem Res 11:1261–1272.

Cajal SR (1901–1917) Recollections of My Life (Horne Craigie, E, trans.), 1st MIT Press, paperback ed., Cambridge, MA, 1989.

Cammermeyer J (1966) Morphologic distinctions between oligodendrocytes and microglia cells in the rabbit cerebral cortex. Am J Anat 118: 227–248.

Cardone B Roots BI (1990) Characterization of the myelin-like membranes in earthworm CNS using immunolabelling techniques. Soc Neurosci Abstr 16:666.

Cardone B Roots BI (1991) Comparative studies of myelin-like membranes in annelids and arthropods. Trans Am Soc Neurochem 22:158.

Choi BH Kim RC (1985) Expression of glial fibrillary acidic protein by immature oligodendroglia and its implications. J Neuroimmunol 8:215–235.

Choi BH Kim RC Lapham LW (1983) Do radial glia give rise to both astroglial and oligodendroglial cells? Dev Brain Res 8:119–130.

Coggeshall RE (1965) A fine structural analysis of the ventral nerve cord and associated sheath of *Lumbricus terrestris* L. J Comp Neurol 125:393–437.

Cuadras J Marti-Subirana A (1987) Glial cells of the crayfish and their relationships with neurons. An ultrastructural study. J Physiologie Paris 82:196–217.

Drewes CD Brinkhurst RO (1990) Giant nerve fibers and rapid escape reflexes in newly hatched aquatic oligochaetes, *Lumbriculus variegatus* (Family Lumbriculidae). Invert Reprod Devel 17:91–95.

Dubois-Dalcq M (1987) Characterization of a slowly proliferating cell along the oligodendrocyte differentiation pathway. EMBOJ 6:2587–2595.

Dubois-Dalcq M Behar T Hudson L Lazzarini RA (1986) Emergence of three myelin proteins in oligodendrocytes cultured without neurons. J Cell Biol 102:384–392.

Espinosa de los Monteros A Zhang M De Vellis J (1993) 02A progenitor cells transplanted into the neonatal rat brain develop into oligodendrocytes but not astrocytes. Proc Natl Acad Sci USA 90:50–54.

Fernandez-Valle C Fregien N Wood PM Bunge MB (1993) Expression of the protein zero myelin gene in axon-related Schwann cells is linked to basal lamina formation. Development 119:867–880.

Fraher JP (1972) A quantitative study of anterior root fibres during early myelination. J Anat 112:99–124.

Fraher JP (1976) The growth and myelination of central and peripheral segments of ventral motoneurone axons. A quantitative ultrastructural study. Brain Res 105:193–211.

Goldman JE (1992) Regulation of oligodendrocyte differentiation. TINS 15: 359–362.

Gould RM Jessen KR Mirsky R Tennekoon G (1992) The Cell of Schwann: An update, in Myelin: Biology and Chemistry (Martenson RE, ed.), CRC Press, Boca Raton, FL, pp 123–171.

Günther J (1973) A new type of "node" in the myelin sheath of an invertebrate nerve fibre. Experientia 29:1263–1265.

Günther J (1976) Impulse conduction in the myelinated giant fibers of the earthworm. Structure and function of the dorsal nodes in the median giant fiber. J Comp Neurol 168:505–531.

Hahn AF Chang Y Webster H de F (1987) Development of myelinated nerve fibers in the sixth cranial nerve of the rat: a quantitative electron microscope study. J Comp Neurol 260:491–500.

Hama K (1959) Some observations on the fine structure of the giant nerve fibers of the earthworm *Eisenia foetida*. J Cell Biophys Biochem Cytol 6:61–66.

Hama K (1966) The fine structure of the Schwann cell sheath of the nerve fiber in the shrimp *(Penaeus japonicus)*. J Cell Biol 31:624–632.

Hardy R Reynolds R (1991) Proliferation and differentiation potential of rat forebrain oligodendroglial progenitors both *in vitro* and *in vivo*. Development 111:1061–1080.

Hess A (1965) Developmental changes in the structure of the synapse on the myelinated cell bodies of the chicken ciliary ganglia. J Cell Biol 25:1–19.

Heuser JE Doggenweiler CF (1966) The fine structural organization of nerve fibres, sheaths and glial cells in the prawn *Palaemonetes vulgaris*. J Cell Biol 30:381–403.

Holmes W (1942) The giant myelinated nerve fibres of the prawn. Phil Trans R Soc Lond (B) 231:293–311.

Holtzman FA Freeman AR Kashner LA (1970) A cytochemical electron microscope study of channels in the Schwann cells surrounding lobster giant axons. J Cell. Biol 44:438–444.

Hörstadius S (1950) The Neural Crest, Oxford University Press, London, p. 111.

Hudson LD Nadon NL (1992) Amino acid substitutions in proteolipid protein that cause dysmyelination, in Myelin: Biology and Chemistry (Martenson RE, ed.), CRC Press, Boca Raton, FL, pp. 677–702.

Jeserich G Rauen T (1990) Cell cultures enriched in oligodendrocytes from the central nervous system of trout in terms of phenotypic expression exhibit parallels with cultured rat Schwann cells. Glia 3:65–74.

Jeserich G Stratmann A (1992) *In vitro* differentiation of trout oligodendrocytes: evidence for an A2B5-positive origin. Dev Brain Res 67:27–35.

Jeserich G Waehneldt TV (1986) Bony fish myelin: Evidence for common major structural glycoproteins in central and peripheral myelin of trout. J Neurochem 46:525–533.

Jeserich G Waehneldt-Kreysing TVB (1992) An evolutionary approach to myelin proteins and myelin-forming cells in the vertebrate brain. Biochem Soc Trans 20:617–620.

Jeserich G Müller A Jacque C (1990) Developmental expression of myelin oligodendrocytes in the CNS of trout. Dev Brain Res 51:27–34.

Jessen KR Mirsky R (1991) Schwann cell precursors and their development. Glia 4:185–194.

Jessen KR Brennan A Morgan L Mirsky R Kent A Hashimoto Y Gavrilovic J (1994) The Schwann cell precursor and its fate: A study of cell death and differentiation during gliogenesis in rat embryonic nerves. Neuron 12:1–20.

Johnson GE (1924) Giant nerve fibers in crustaceans with special reference to *Cambarus* and *Palaemonetes*. J Comp Neurol 36:323–373.

Kirschner DA Blaurock AE (1992) Organization, phylogenetic variations and dynamic transitions of myelin, in Myelin: Biology and Chemistry (Martenson RE, ed.), CRC Press, Boca Raton, FL, pp. 3–78.

Kitagawa K Fidler L Sinoway MP Yang CW Hashim G Colman DR (1993) The myelin proteolipid proteins: evolution of the gene family in lower vertebrates. Trans Am Soc Neurochem 24:228.

Kruger L Maxwell DS (1967) Comparative fine structure of vertebrate neuroglia: teleosts and reptiles. J Comp Neurol 129:115–142.

Landon DN Hall S (1976) The myelinated nerve fibre, in The Peripheral Nerve (Landon DN, ed.), Chapman and Hall, London, pp. 1–105.

Lane NJ Abbott NJ (1975) The organization of the nervous system in the crayfish *Procambarus clarkii*, with emphasis on the blood-brain interface. Cell Tiss Res 156:173–187.

Le Douarin NM (1982) The Neural Crest. Cambridge University Press, Cambridge.

Le Douarin NM Dulac C Dupin E Cameron-Curry P (1991) Glial cell lineages in the neural crest. Glia 4:175–184.

Lubetzki C Goujet-Zalc C Gansmüller A Monge M Brillat A Zalc B (1991) Morphological biochemical and functional characterization of bulk isolated glial progenitor cells. J Neurochem 56:671–680.

Lubetzki C Demerens C Anglade P Villarroya H Frankfurter A Lee V MY Zalc B (1993) Even in culture, oligodendroctyes myelinate solely axons. Proc Natl Acad Sci USA 90:6280–6284.

Lyser KM (1972) The fine structure of glial cells in the chicken. J Comp Neurol 146:83–94.

Matthews MA Duncan D (1971) A quantitative study of the morphological changes accompanying the initation and progress of myelin production in the dorsal funiculus of the rat spinal cord. J Comp Neurol 142:1–22.

McAlear JH Milburn NS Chapman GB (1958) The fine structure of Schwann cells, nodes of Ranvier and Schmidt-Lantermann incisures in the central nervous system of the crab, *Cancer irroratus*. J Ultrastruct Res 2:171–176.

Mirksy R Winter J Abney ER Pruss RM Gavrilovic J Raff MC (1980) Myelin specific proteins and glycolipids in rat Schwann cells and oligodendrocytes in cultures. J Cell Biol 84:483–494.

Monzon-Mayor M Yanes C James JL Sturrock RR (1990) An ultrastructural study of the development of oligodendrocytes in the midbrain of the lizard. J Anat 170:43–49.

Mori S Leblond C (1970) Electron microscopic identification of three classes of oligodendrocytes and a preliminary study of their proliferative activity in the corpus callosum of young rats. J Comp Neurol 139:1–30.

Mugnaini E Walberg F (1964) Ultrastructure of neuroglia. Ergebn. Anat Entw Gesch 37:194–236.

Nageotte J (1916) Note sur les fibres à myéline et sur les étranglements de Ranvier chez certains crustacés. CR Soc Biol Paris 79:259–263.

Nicol JAC (1948) The giant axons of annelids. Q Rev Biol 23:291–323.

Noll E Miller RH (1993) Oligodendrocyte precursors originate at the ventral ventricular zone dorsal to the ventral midline region in the embryonic rat spinal cord. Development 118:563–573.

Nona SN Duncan A Stafford CA Maggs A Jeserich G Cronly-Dillon JR (1992) Myelination of regenerated axons in goldfish optic nerve by Schwann cells. J Neurocytol 21:391–401.

Norlander RH Singer M (1972) Electron microscopy of severed motor fibers in the crayfish. Z Zellforsch, 126:157–181.

Ochoa J (1976) The unmyelinated nerve fibre, in The Peripheral Nerve (Landon DN, ed.), Chapman and Hall, London, pp. 106–158.

Okamura N Stoskopf M Yamaguchi H Kishimoto Y (1985a) Lipid composition of the nervous system of earthworms *(Lumbricus terrestris)*. J Neurochem 45:1875–1879.

Okamura N Stoskopf M Hendricks F Kishimoto Y (1985b) Phylogenetic dichotomy of nerve glycosphingolipids. Proc Natl Acad Sci USA 82:6779–6782.

Okamura N Yamaguchi H Stoskopf M Kishimoto Y Saida T (1986) Isolation and characterization of mutlilayered sheath membrane rich in glucocerebroside from shrimp ventral nerve. J Neurochem 47:1111–1116.

Ona A (1993) The mammalian vestibular ganglion cells and the myelin sheath surrounding them. Acta Otolaryngol (Stokh) Suppl 503:143–149.

Parnavelas JG Luder R Pollard SG Sullivan K Lieberman AR (1983) A qualitative and quantitative ultrastructural study of glial cells in the developing visual cortex of the rat. Phil Trans Roy Soc Lond B 301:55–84.

Peracchia C Robertson JD (1971) Increase in osmophilia of axonal membranes of crayfish as a result of electrical stimulation, asphyxia, or treatment with reducing agents. J Cell Biol 51:223–239.

Pereyra P Roots BI (1988) Isolation and initial characterization of myelin-like membrane fractions from the nerve cord of earthworms *(Lumbricus terrestris).* Neurochem Res 13:893–901.

Peters A Palay SL Webster H de F (1991) The Fine Structure of the Nervous System. 3rd ed., Oxford University Press, New York, p. 494.

Pinching AJ (1971) Myelinated dendritic segments in the monkey olfactory bulb. Brain Res 29:133–138.

Pringle NP Richardson WD (1993) A singularity of PDGF alpha-receptor expression in the dorsoventral axis of the neural tube may define the origin of the oligodendrocyte lineage. Development 117:525–533.

Privat A (1975) Postnatal gliogenesis in the mammalian brain. Int Rev Cytol 40:281–323.

Prosser CL (1961) In Comparative Animal Physiology (Prosser CL Brown FA, eds.), Saunders, Philadelphia, 2nd ed., p. 595.

Raff MC Miller RH Noble M (1983) A glial progenitor cell that develops *in vitro* into an astrocyte or an oligodendroctye depending on culture medium. Nature 303:390–396.

Raine CS (1984) Morphology of myelin and myelination, in Myelin (Morell P, ed.), 2nd ed., Plenum, New York, pp. 1–50.

Remahl S Hildebrand C (1982) Changing relation between onset of myelination and axon diameter range in developing feline white matter. J Neurol Sci 54:33–45.

Retzius G (1890) Zur Kenntniss des Nervensystems der Crustaceen. Biol Untersuch n F 1:1–50.

Reynolds R Wilkin GP (1988) Development of macroglial cells in rat cerebellum II. An *in situ* immunohistochemical study of oligodendroglial lineage from precursor to mature myelinating cell. Development 102:490–425.

Reynolds R Wilkin GP (1991) Oligodendroglial progenitor cells but not oligodendroglia divide during normal development of the rat cerebellum. J Neurocytol 20:216–224.

Roots BI (1978) A phylogenetic approach to the anatomy of glia, in Dynamic Properties of Glia Cells (Schofeniels E Franck B Hertz L Towers DB, eds.), Pergamon, New York, pp. 45–54.

Roots BI (1984) Evolutional aspects of the structure and function of the nodes of Ranvier, in The Node of Ranvier (Zagoren JC Fedoroff S, ed.), Academic, Orlando, FL, pp. 1–29.

Roots BI (1993) The evolution of myelin, in Advances in Neural Sciences, vol. 1 (Malhotra S, ed.), JAI Press, Greenwich, CT, pp. 187–213.

Roots BI Lane NJ (1983) Myelinating glia of earthworm giant axons: Thermally induced intramembranous changes. Tiss Cell 15:695–709.

Rosenbluth J (1962) The fine structure of acoustic ganglia in the rat. J Cell Biol 12:329–359.

Rosenbluth J Palay SL (1961) The fine structure of nerve cell bodies and their myelin sheaths in the eighth nerve ganglia of the goldfish. J Biophys Biochem Cytol 9:853–877.

Schnapp B Mugnaini E (1975) The myelin sheath: electron microscopic studies with thin sections and freeze-fracture, in Golgi Centennial Symposium Proceedings (Santini M, ed.), Raven, New York, pp. 209–233.

Schonbach C (1969) The neuroglia in the spinal cord of the newt *Triturus viridescens*. J Comp Neurol 135:93–120.

Shivers R Brightman MW (1976) Transglial channels in ventral nerve roots of crayfish. J Comp Neurol 167:1–26.

Sivron T Jeserich G Nona S Schwartz M (1992) Characteristics of fish glial cells in culture: Possible implications as to their lineage. Glia 6:52–66.

Skoff RP (1980) Neuroglia: A re-evaluation of their origin and development. Path Res Pract 168:279–300.

Stensaas LJ (1977) The ultrastructure of astrocytes, oligodendrocytes and microglia in the optic nerve of urodele amphibians *(A. punctatum T. pyrrhogaster T. viridescens)*. J Neurocytol 6:269–286.

Stensaas LJ Stensaas SS (1968a) Astrocytic neuroglial cells, oligodendrocytes and microgliacytes in the spinal cord of the toad. I Light microscopy. Z Zellforsch 84:473–489.

Stensaas LJ Stensaas SS (1968b) Astrocytic neuroglial cells, oligodendrocytes and microgliacytes in the spinal cord of the toad. II Electron microscopy. Z Zellforsch 86:184–213.

Tamai Y Kojima A Saito S Takayama-Abe K Horichi H (1992) Characteristic distribution of glycolipids in gadoid fish nerve tissues and its bearing on phylogeny. J Lipid Res 33:1351–1359.

Tamsit SG Bally-Cuif L Colman DR Zalc B (1992) DM-20 mRNA is expressed during the embryonic development of the nervous system of the mouse. J Neurochem 58:1172–1175.

Waehneldt TV (1990) Phylogeny of myelin proteins, in Myelination and Dysmyelination, vol. 605, Ann NY Acad Sci, pp. 15–28.

Waehneldt TV Malotka J Kitamura S Kishimoto Y (1989) Electrophoretic characterization and immunoblot analysis of the proteins from the myelin-like light membrane fraction of shrimp ventral nerve *(Penaeus duorarum)*. Comp Biochem Physiol 92B:369–374.

Warf BC Fok-Seang J Miller RH (1991) Evidence for the ventral origin of oligodendrocyte precursors in the rat spinal cord. J Neurosci 11:2477–2488.

Warren RH Rubin RW (1978) Microtubules and actin in giant nerve fibers of the spiny lobster *Panuliris argus*. Tissue Cell, 11:487–697.

Warrington AE Pfeiffer SE (1992) Proliferation and differentiation of 04^+ oligodendrocytes in postnatal rat cerebellum: analysis in unfixed tissue slices using anti-glycolipid antibodies. J Neurosci Res 33:338–353.

Willey TJ (1973) The ultrastructure of the cat olfactory bulb. J Comp Neurol 152:211–232.

Williams BP Price J (1992) What have tissue culture studies told us about the development of oligodendrocytes? Bioessays 14:693–698.

Wilson EB (1889) The embryology of the earthworm. J Morph 3:387–462.

Wolburg H Bouzehouane U (1986) Comparison of the glial investment of normal and regenerating fiber bundles in the optic nerve and optic tectum of the goldfish and the Crucian carp. Cell Tissue Res 244:187–192.

Zoran MJ Drewes CD Fourtner CR Siegel AJ (1988) The lateral giant fibers of the tubificid worm *Branchiura sowerbyi*: Structural and functional asymmetry in a paired interneuronal system. J Comp. Neurol 275:76–86.

A Cellular and Molecular Approach
to Myelinogenesis
in the CNS of Trout

Gunnar Jeserich,
Astrid Stratmann, and Jens Strelau

1. Fundamentals of Myelinogenesis
in Higher Vertebrates

Myelin is an extension of the glial plasmamembrane that surrounds axons as a multilamellar insulating sheath to permit a most efficient and fast mode of signal propagation, the saltatory impulse conduction (Raine, 1984; Ritchie, 1984). Two distinct types of glial cells are responsible for the elaboration of myelin in central and peripheral nerve fibers, respectively, which basically differ from each other in terms of cell structure, molecular phenotype, and cell lineage relationship. Oligodendrocytes, the myelin-forming cells in the central nervous system (CNS) each elaborate numerous slender processes to ensheath a multitude of internodes on different axons (Sternberger et al., 1978; Wood and Bunge, 1984; Pfeiffer et al., 1993). Schwann cells, which produce myelin in the peripheral nervous system (PNS) on the other hand, are always closely associated with an individual nerve fiber and strictly establish a one-to-one relationship with a single internodal segment (Peters et al., 1976; Rumsby and Crang, 1977); Furthermore, these cells typically are surrounded by a basal lamina produced by themselves, which is critically involved in the initiation of the myelination process in the PNS (Bunge et al., 1986). This diversity in cellular morphology further extends into characteristic dif-

From: *Neuron-Glia Interrelations During Phylogeny: I. Phylogeny and Ontogeny of Glial Cells*
A. Vernadakis and B. Roots, Eds. Humana Press Inc., Totowa, NJ

ferences in the biochemical features of the myelin synthesized by either cell type (Fig. 1). Myelin generated in the CNS by oligodendrocytes is characterized by a high proportion of a strongly hydrophobic, nonglycosylated proteolipidprotein (PLP) (Lees and Brostoff, 1984), which comprises four transmembrane segments and is thought to accomplish the tight apposition of external glial membrane faces (Laursen et al., 1984; Stoffel et al., 1984) at the ultrastructural level giving rise to the intraperiod line (Braun, 1984). The major gene product of Schwann cells by contrast is a hydrophobic glycoprotein of about 29 kDa mol wt, possessing only a single membrane spanning domain, which is called P_0 (Everly et al., 1973; Lemke and Axel, 1985). The extracellular domain of this protein exhibits an immunoglobulin-like secondary structure enabling it to convey external adhesion of the glial lamellae by homophilic molecular interactions (Lemke, 1988; Filbin et al., 1990; D'Urso et al., 1990). The cytoplasmic portion of the molecule is highly basic and is thought to align with polar headgroups of membrane lipids, thus helping to stick together adjacent cytoplasmic membrane faces (Lemke and Axel, 1985). Although PLP and P_0 seem to functionally replace each other in CNS and PNS myelin, respectively, they do not exhibit any appreciable sequence homologies and are entirely different with respect to their membrane topology and genomic arrangements (Laursen et al., 1984; Stoffel et al., 1984; Lemke et al., 1988; Nave and Milner, 1989). A major structural constituent shared by both types of myelin is myelin-basic protein (Carnegie and Dunkley, 1975; Roach et al., 1983), a hydrophilic component of about 18 kDa with an isoelectric point of about 10, which occurs in a number of alternative splice variants in the CNS (deFerra et al., 1985). All of them are extrinsic myelin components synthesized on free polyribosomes (Colman et al., 1982), their mRNAs being transported out into the cellular periphery along cytoskeletal tracts (Brophy et al., 1993). The translated proteins are attached to the cytoplasmic surface of the glial extensions to fuse myelin at the major dense line (Privat et al., 1987; Omlin et al., 1988). Compact myelin as described above is a feature unique to the vertebrate nervous tissue that is initially encountered in the phylogenetically ancient classes of fish (bony and cartilaginous fishes), whereas it does not occur in invertebrates (*see also* previous chapter).

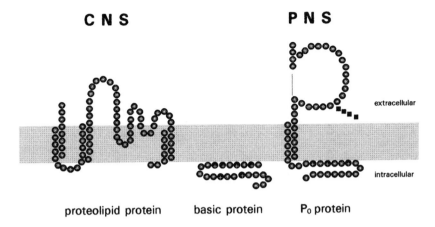

Fig. 1. Molecular model and membrane topology of major myelin proteins of mammalian CNS and PNS, respectively.

2. Properties of Myelin Proteins and Myelin-Forming Cells in Bony Fish

In bony fish, as in higher vertebrates, two different types of myelin-forming cells can be distinguished by their morphology: according to electronmicroscopical analyses (Kruger and Maxwell, 1967; Jeserich and Waehneldt, 1986a) as well as immunocytochemical studies. Glial cells in myelinated fiber tracts of the teleost CNS exhibit a close structural similarity to mammalian oligodendrocytes in that they extend multiple slender processes to myelinating fibers in their vicinity (Fig. 2). In the lateral line nerve of trout, which is part of the peripheral nervous system, on the other hand, all the characteristic features of Schwann cell myelination are found to be indistinguishable from those of the mammalian PNS described earlier. The ultrastructure of the myelin sheath as well shows virtually the same structural organization of alternating dense lines as seen in the mammalian CNS and PNS, respectively (Jeserich and Waehneldt, 1986a).

A comparative biochemical analysis of myelin isolated from the CNS and PNS of trout by SDS-PAGE and Western blotting, however, furnished a quite exceptional protein composition. As shown by Fig. 3, the protein profile of trout CNS myelin comprises four major components of about equal amount: a fastly migrating component of 13 kDa apparent mol wt, which strongly

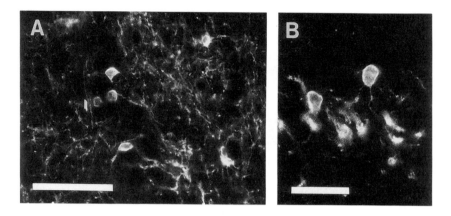

Fig. 2. Immunohistochemical localization of 36 kDa protein in oligo-dendrocytes and their processes in trout spinal cord (**A**) and optic tectum (**B**). Note diffuse cytoplasmic staining. Scale bars indicate 45 µm in A and 23 µm in B. Reproduced with permission from Jeserich and Waehneldt, 1986b.

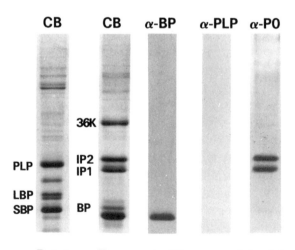

Fig. 3: SDS-PAGE of myelin proteins from the CNS of rat (lane 1) and trout (lanes 2–5). CB: Coomassie blue staining, α-BP: antihuman myelin basic protein immunoblot, α-PLP: antibovine proteolipid protein immunoblot, α-P_0: antibovine P_0 immunoblot.

crossreacts with antibodies against mammalian myelin basic protein and on 2D gels appears slightly less basic than its mammalian counterpart. A similar protein has been identified in goldfish myelin as well (Roots et al., 1984). Furthermore, two peptides of intermediate molecular size (termed IP1: 23 kDa and IP2: 26 kDa) on the gel running close to the mammalian proteolipid protein are found. Probing the blots with anti-PLP antibodies, however, did not give any perceptible signal, suggesting that an immunologically related component does not exist in fish CNS myelin (Waehneldt and Jeserich, 1984). The two intermediately sized proteins instead proved as concanavalin A binding glycoproteins strongly crossreacting with antibodies against the major PNS myelin glycoprotein P_0 (Jeserich and Waehneldt, 1986a). Since the staining persists after enzymatic deglycosylation, it can be attributed to sequence homologies in the amino acid chains and is not reflecting common epitopes in the oligosaccharide portion (Jeserich and Waehneldt, 1987). Recent cDNA sequence analysis of the trout IP1 isoform confirmed its structural relatedness with mammalian P_0 (Stratmann and Jeserich, 1995). The degree of sequence identity with rat P_0 was highest in the extracellular domain (about 50%) and significantly lower in the remaining portions of the molecule. Most important, the position of the two cysteine residues required for stabilization of the immunoglobulin-fold by disulfide bridging is exactly conserved in both cases. The same is true for the putative glycosylation site at Asn^{93}. In the cytoplasmic portion IP1 is shorter by 38 amino acids than mammalian P_0, explaining the differences in the electrophoretic mobility. Furthermore, this region shows only little sequence conservation, albeit being characterized by a rather high proportion of basic residues in both cases. The molecular structure of the larger isoform (IP2) is still unknown so far. Results of controlled autolysis experiments as well as limited proteolysis studies in combination with Western blot analysis, however, suggest that both proteins are highly homologous (Waehneldt and Jeserich, 1984; Jeserich and Waehneldt, 1986a). Since the difference in molecular size is retained after enzymatic deglycosylation it cannot be accounted for by variations in the oligosaccharide moiety but has to be owing to deviations in the amino acid chain. Immunocytochemical studies on trout oligodendrocytes with a monoclonal antibody that selectively recognizes the IP1 molecule but not IP2 reveal that the distinguishing epitopes are not extracellularly located but seem to reside in the cytoplasmic region. This would also agree with the results of 2D gel analysis, demonstrating that IP2 is significantly more basic than IP1, probably owing to a larger basic cytoplasmic tail.

Another striking biochemical feature of fish CNS myelin is the occurrence of a novel unglycosylated component of about 36 kDa apparent mol wt (Jeserich, 1983), termed 36K, which does not crossreact with antibodies against any of the known myelin proteins of higher vertebrate species. Polyclonal antibodies raised against this component on Western blots selectively label a band of the expected electrophoretic mobility in the CNS myelin of all bony fish species thus far examined (Waehneldt et al., 1986b), but fail to recognize any protein in the myelin of other vertebrate classes including mammals, birds, reptiles, and amphibia. Furthermore, recent nucleotide sequence analysis of the cloned 36 kDa, cDNA has revealed virtually no structural similarity to other sequenced proteins, confirming the uniqueness of this component. Hydrophobicity analysis of the deduced amino acid sequence does not identify a clear membrane spanning domain, suggesting that 36 kDa is not an integral protein but rather is attached to the inner cytoplasmic surface of the myelin membrane, a site where also myelin basic protein is known to reside. It is interesting to note in this context that in vitro metabolic labeling experiments performed with the trout optic nerve have revealed a similar kinetics of entry into myelin membranes for both MBP and 36 kDa. Furthermore, immunocytochemical studies on the light and electron microscopical level have consistently revealed a homogenous distribution of 36 kDa in the cytoplasm of oligodendrocytes as it is generally characteristic for proteins synthesized on free polyribosomes. It would be interesting to determine more directly the site of synthesis of these myelin proteins in fish by an in vitro translation approach using rabbit reticulocyte lysates programmed with polyribosomal mRNA isolated from the trout CNS.

Immunocytochemical staining, furthermore, selectively localizes 36 kDa to the myelin sheaths of CNS fiber tracts, although it is not detectable in the PNS (Jeserich and Waehneldt, 1986b). Since this is confirmed by comparative electrophoretic protein analysis, it is clear that the protein is unique to CNS myelin membranes. The functional purpose of this novel major myelin constituent is still an enigma. The fact that it occurs without exception throughout the huge group of bony fishes implies that it is of crucial functional importance.

As a whole, the myelin-forming cells in the brain of fish in terms of morphology closely match oligodendrocytes of the mammalian CNS, whereas in their biochemical properties they strikingly resemble mammalian Schwann cells, in that they express two P_0-like proteins but no proteolipid protein. A closer characterization of physiological properties of this particular type of oligodendrocyte therefore was of obvious interest.

3. Developmental Expression of Myelin Proteins in the Trout CNS

3.1. Immunohistochemical Studies

Myelinogenesis in the mammalian brain is a multistep process initiated by the induction of myelin-specific genes in oligodendrocytes followed by the accumulation of corresponding translational products that are subsequently transported out along the cytoplasmic processes and incorporated into myelin membranes. Already the primary event, the functional differentiation of oligodendrocytes proceeds in a highly coordinated fashion in that the myelin-specific constituents are expressed sequentially and on an ordered schedule that is predictable both in time and space. In the rodent brain the appearance of galactocerobroside and 2',3'-cyclic nucleotide 3'-phosphohydrolase (CNP) denotes the initial step of oligodendrocyte maturation, followed by the expression of myelin-basic protein and, finally, proteolipidprotein (Monge et al., 1986). A similarly ordered progression of oligodendrocyte maturation is observed in the developing fish brain as well (Fig. 4). Using a panel of monoclonal and polyclonal antibodies raised against the major protein constituents of trout CNS myelin the molecular differentiation of oligodendrocytes could be monitored. By immunohistochemical staining of freshly dissociated cells from total brains of trout at various developmental ages, a constant chronological sequence of myelin protein expression by oligodendrocytes was revealed (Jeserich et al., 1990a): Glycoprotein IP2 together with galactocerebroside were the earliest myelin markers to appear during glial differentiation at stage 28 (1 wk before hatching, Vernier, 1969) being followed by the

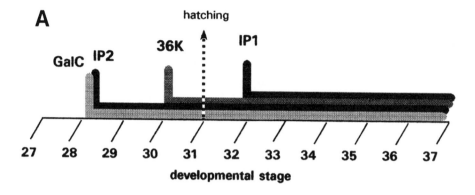

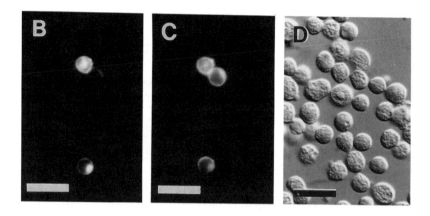

Fig. 4. Expression of myelin constituents by oligodendrocytes during larval development of trout brain. **(A)** Schematic drawing illustrating the sequential appearance of myelin antigens. GalC. Galactocerebroside. IP1, IP2, and 36 kDa: major trout CNS myelin proteins. **(B–D)** Freshly dissociated cell preparations derived from trout brains at developmental stage 34 double-labeled with monoclonal antibodies against IP1 (B) in combination with anti-36 kDa antiserum (C). D: Nomarski interferential contrast photomicrograph. Scale bars indicate 19 μm. B–D reproduced with permission from Jeserich et al., 1990.

occurrence of 36 kDa at stage 30 (around hatching) and finally glycoprotein IP1 at stage 32.

This sequence of myelin protein expression was confirmed by double-labeling experiments using a monoclonal antibody in conjunction with a polyclonal antiserum tagged with different fluorophores. In cell preparations derived from trout brains at stage 32, the great majority of immunolabeled cells selectively stained with anti-IP2 antibodies and only a minor proportion of them exhibited double-labeling for both myelin glycoproteins, whereas cells staining for IP1 alone did not occur at any age. Similarly, in double-labeling experiments using an anti-36 kDa antiserum in combination with the monoclonal antibody 7C4 (which selectively recognizes the IP1 molecule, Jeserich et al., 1990b), the cells exhibited either a 36 kDa+/IP1- or, more frequently, a mixed 36 kDa+/IP1+ phenotype, whereas IP1+/36 kDa- cells were not encountered, emphasizing that expression of 36 kDa precedes those of IP1 during functional differentiation of trout oligodendrocytes (Fig. 4).

The finding that the two P_0-like glycoproteins, IP2 and IP1, are sequentially expressed is consistent with comparative electrophoretic data, demonstrating that the proportion of IP1 is low in myelin from early stages of brain development and continuously increases during myelin maturation (Jeserich, 1983). Therefore, it is tempting to speculate that IP1 specifically helps to stabilize the myelin sheath during myelin compaction. The structural organization of the extracellular domain of IP1 as an adhesion molecule would certainly fit into such a concept.

Moreover, the observation that GalC in fish as in the mammalian brain represents one of the earliest markers of oligodendrocyte differentiation suggests that this glycolipid might play a key role in this process. In fact, recent antibody perturbation experiments done with cultured rat oligodendrocytes (Bansal and Pfeiffer, 1989) demonstrated that anti-GalC antibodies reversibly block the differentiation of immature oligodendrocytes being reflected in terms of cellular morphology as well as antigenic expression. Since anti-GalC treatment furthermore elicits an influx of calcium ions into the cells, a role for these glycolipoids in transmembrane signaling affecting the organization of the cytoskeleton has been suggested (Dyer and Benjamins, 1990). It would be interesting to establish, whether corresponding effects can be observed in fish oligodendrocytes, too.

Immunohistochemical staining of frozen tissue sections through the brain of trout corroborated the earlier described pattern of antigenic expression on the level of the nascent myelin sheath and gave insight into the onset and spatio-temporal course of myelinogenesis in the various brain regions. Thus, it was demonstrated that in the CNS of trout myelinogenesis is initiated in the spinal cord and hindbrain, whereas in the tectum and cerebellum it starts significantly later, being in full accordance with the results of electron microscopical investigations (Jeserich, 1981). Interestingly, the first myelinating fibers in the spinal cord and medulla oblongata were typically arranged in two major fiber tracts of the anterior funiculus region being aligned ventrally on either side of the central canal and probably included descending motor fibers regulating tail and trunk muscle activity. Myelination of these fiber tracts clearly coincided with the ability of the larvae for tail flip movements and initial swimming activity. On longitudinal sections these fiber tracts could be further traced into the anterior regions of the brainstem. A similar correlation between myelination of a certain nerve fiber tract and the accretion of functional responsiveness could be monitored in the trout visual pathway.

3.2. In Situ *Hybridization*

In order to follow myelin gene expression at the level of nucleic acids the cDNA coding for the 36 kDa protein of trout was subcloned into a pGEM4 vector to generate sense and antisense RNA probes for *in situ* hybridization. On northern blots of total trout brain mRNA, a single band of about 2 kb was selectively labeled, whereas with trout liver mRNA no reaction was observed at all. On tissue sections through the mature trout brain the 36K transcript was selectively localized to cells with an oligodendroglial morphology in myelinated fiber tracts, being most clearly visible in the visual pathway (Fig. 5). Unmyelinated brain regions as well as control sections treated with a sense RNA probe were not labeled.

During development the 36 kDa mRNA initially occurred in a defined area of the medulla oblongata and spinal cord (anterior funiculus) shortly before hatching from the egg, i.e., at the same site where myelin deposition is initially encountered, whereas in higher brain regions, such as the visual pathway, it emerged significantly later. In a given brain area, immunostaining for the 36 kDa protein was always closely correlated with the occurrence of its respective transcript, suggesting that expression of this myelin component is regulated at the level of nucleic acids and probably is under transcriptional control.

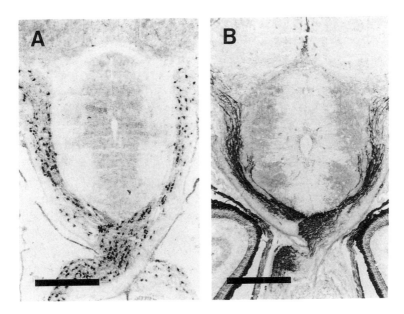

Fig. 5. Transverse sections through the midbrain of trout at developmental stage 35 exhibiting the visual pathway. **(A)** *In situ* hybridization localization of the 36 kDa transcript in oligodendrocytes using a digoxigenin-labeled antisense mRNA probe transcribed from trout 36 kDa cDNA. **(B)** Adjacent section showing immunohistochemical localization of the 36 kDa protein in myelinated fiber tracts. Scale bars indicate 270 μm.

4. Regulation of Myelin Gene Expression in Trout Oligodendrocytes: Studies on Cell Cultures

Mammalian oligodendrocytes and Schwann cells basically differ from each other regarding the regulation of myelin gene expression, in particular with respect to their dependence on a persistent contact with neurons to elaborate correctly their developmental program (Mirksy et al., 1980). Thus, immature oligodendrocytes or their precursor cells are able to go through a complete sequence of biochemical differentiation when cultured in the entire absence of neurons, and develop under these conditions in the same time frame as they do in vivo (Abney et al., 1981; Zeller et al., 1985; Gard and Pfeiffer, 1989). Schwann cells on the other hand strictly require an ongoing axonal contact for proper induction and maintenance of their myelinogenic phenotype during development and regeneration and hence rapidly downregulate their myelin-related molecules on axonal withdrawal in dissociated cell culture or after nerve transection in vivo (Brockes et al., 1979; Lemke and Chao, 1988; Trapp et al., 1988).

Since fish oligodendrocytes, as mentioned earlier, seem to share characteristic features with both types of glial cells the regulation of myelin gene expression in this particular type of myelin-forming cell appeared of special interest. To examine whether trout oligodendrocytes required a permanent interaction with neurons for a normal progression of cell differentiation, dissociated glial cultures were prepared from trout brains at different developmental ages and the expression of myelin markers was monitored immunohistochemically after various periods in vitro (Jeserich and Stratmann, 1992). In principle two different types of cells developed in these cultures. One exhibited a flattened polygonal cell shape reminiscent of type 1 astrocytes as seen in mammalian glial cultures and was GFAP positive. These astrocytes were actively dividing in vitro and soon formed colonies of increasing size. The second cell type was of a multiple branched morphology resembling cultured mammalian oligodendrocytes and selectively stained with antibodies against myelin proteins. Since the glycoprotein IP2 is the first myelin marker to occur during oligodendrocyte differentiation in vivo, the monoclonal antibody 6D2 recognizing an epitope in the carbohydrate chain of this protein was used to identify oligodendrocytes in culture and to define their antigenic phenotype by dual labeling immunostaining. In glial cultures prepared from larval stage 30 immediately after plating, only a rather low proportion of cells expressed this myelin marker. During the following 9 d in vitro, the percentage of IP2+ cells increased twofold, attaining a similar level as previously determined in the larval trout brain *in situ* at the equivalent in vivo age. Since IP2+ cells did not divide in culture this increase can only be explained by a continuous conversion of immature progenitor cells into committed oligodendrocytes.

To examine the further progression of antigenic differentiation in vitro double-labeling experiments using anti-36 kDa antiserum in combination with the monoclonal antibody 6D2 were performed. In cell preparations derived from brains at larval stage 31 (shortly after hatching) the majority of 6D2+ positive oligo-dendrocytes immediately after seeding were still 36 kDa–. Sur-prisingly, during a period of 12 d in vitro these cells did not further differentiate to acquire a 36 kDa+ phenotype, denoting the subse-quent step of oligodendroglial maturation in vivo. Even in glial cultures derived from later developmental stages, which imme-diately after seeding already contained a considerable proportion of 36 kDa+/6D2+ cells expression of 36 kDa was rapidly downregulated to an undetectable level after a few days in cul-ture. During further in vivo maturation trout oligodendrocytes seemed to acquire a certain stabilization of their antigenic pheno-type, since in cultures prepared from stage 34 onward, the decrease of 36 kDa expression proceeded much more slowly. A similar correlation was observed in the case of IP1 expression, reflecting the terminal differentiation step *in situ*. Double-label-ing experiments using the monoclonal antibody 7C4, which selectively labels the IP1 glycoprotein, in combination with anti-IP2 antiserum revealed that immature oligodendrocytes derived from early larval stages did not differentiate in vitro up to the level of IP1 expression. Even those cells having in vivo acquired an IP1+/IP2+ phenotype rapidly ceased expression of IP1 in dis-sociated cell culture. In oligodendrocyte cultures prepared from the mature brain, on the other hand, a spontaneous reinduction of the IP1 glycoprotein was regularly observed after 3 wk in vitro (Jeserich and Rauen, 1990).

To analyze at which metabolic level the expression and downregulation of myelin proteins is controlled in trout oligo-dendrocytes, dot hybridization experiments were performed using antisense RNA probes transcribed from the trout 36 kDa cDNA. Until 2 d in vitro oligodendrocytes from developmental stage 37 expressed high levels of 36 kDa mRNA, whereas by 1 wk, the 36 kDa transcript was no longer perceptible. During the same period in vitro, expression of a housekeeping gene like β-actin did not attenuate. This finding was additionally confirmed by parallel *in situ* hybridization experiments monitoring the expression of the 36 kDa transcript on the level of individual cells. Hence, it is sug-gested that also under in vitro conditions expression of the 36 kDa gene is controlled at the level of nucleic acids, probably by transcriptional mechanisms.

Altogether it is concluded that trout oligodendrocytes, much like mammalian Schwann cells, but in contrast with mammalian oligodendrocytes, require appropriate inducing signals from other cells, e.g., neurons for proper elaboration of their developmental program. At present we do not know the nature of the underlying factors, nor their potential cellular source. The fact that astrocytes were abundant in these cultures makes it unlikely that this cell type was critically involved. It would appear more likely instead that the lack of interactions with neurons was affecting proper maturation of the cells.

5. Oligodendrocyte Progenitor Cells in the Fish CNS

Another distinguishing feature between mammalian Schwann cells and oligodendrocytes pertains to their cellular origin during development. Whereas Schwann cells derive from a population of neural crest cells (Le Douarin, 1986; Jessen and Mirsky, 1991), oligodendrocytes originate in the subventricular zones from bipotential precursor cells that carry the ganglioside epitope A2B5 on their membrane surface (Raff et al., 1983). In order to eventually identify an equivalent progenitor cell type in the developing fish CNS, immunostaining with A2B5 antibodies was performed on freshly dissociated brain cells: In brain tissue from early larval stages (stages 30–34), a rather high proportion of cells in fact exhibited the ganglioside marker and a small part of them were concurrently labeled for the IP2 myelin glycoprotein, suggesting that they were in a transient state of differentiation. As should be expected, the percentage of A2B5+ cells gradually declined with increasing in vivo age but still remained detectable even in the mature brain tissue. Since this indicates that in fish as in mammals, oligodendrocytes originate from A2B5+ progenitor cells, we worked out an immunomagnetic cell separation technique to isolate these cells from the larval trout brain and to follow their developmental fate in culture (Jeserich et al., 1993). The isolation procedure in principle was based on a two-step protocol: The dissociated cells were first centrifuged in a percoll density gradient to yield a glia-enriched cell fraction. From this population, A2B5+ cells could be efficiently separated immunomagnetically by making use of antibody-coated superparamagnetic beads. As revealed by immunostaining, the isolated cell population was highly enriched in A2B5+ cells comprising more than 94% of the whole cell fraction. Immediately after isolation virtually all of these cells were still 6D2–. Within 24 h in culture the A2B5+ cells typically adopted a bipolar or tripolar morphology (Fig. 6), closely resembling oligodendrocyte progenitors of the mammalian brain and were actively proliferating. By 3 d in vitro the majority of these cells had started to express the myelin glycoprotein IP2 and predominantly exhibited a mixed 6D2+/A2B5+ phenotype. By 6 d many of the 6D2+ oligodendrocytes ceased expression of the A2B5 epitope and developed a multipolar shape with numerous fine ramifications, which coincided with the cessation of mitotic activity. Within 10 d in vitro the proportion of 6D2+ oligodendrocytes reached an average of 25%.

In addition to oligodendrocytes a second cell type emerged in the cultures that remained 6D2- but was heavily labeled with anti-goldfish GFAP antibodies (Nona et al., 1989) instead. In terms of morphology, these cells resembled type 1 astrocytes of the mammalian brain and not type 2 as would have been expected from the glial cell lineage in the mammalian optic nerve (Raff et al., 1983). Without exception these astrocytes were A2B5- and hence also in their antigenic phenotype closely matched type 1 astrocytes. It is not entirely clear, however, if both glial cell types appearing in the cultures originated from a common precursor cell or belonged to separate lineages accidentally sharing the A2B5 epitope on their cell surface.

Since in rodent species the glial cell lineage decision of oligodendrocyte progenitor cells is highly susceptible to the amount of serum added to the culture medium we next analyzed the effect of various serum concentrations on the differentiation pathway of glial progenitors in fish: Contrasting with the situation in mammalian optic nerve cultures in trout glial progenitor cells high doses of fetal calf serum (5–10%) induced a twofold increase in the percentage of 6D2+ oligodendrocytes. Switching the cultures from 5 to 0.1% serum after 24 h resulted in a decreased proportion of 6D2+ oligodendrocytes after 6 d. Obviously, this was not owing to an effect on cell proliferation, since the mitotic activity did not change under these conditions. Most surprisingly, neither cell proliferation of trout oligodendrocyte progenitors nor their differentiation into 6D2+-cells were significantly affected by

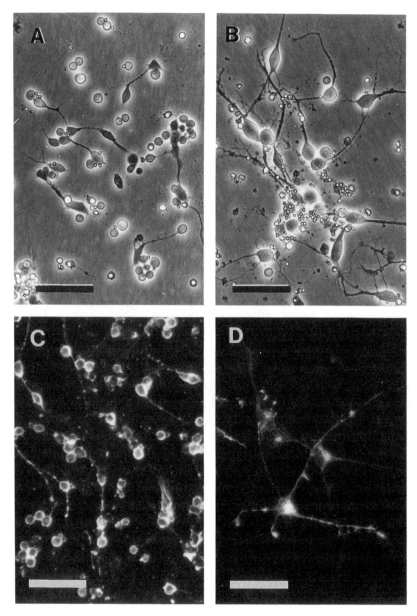

Fig. 6. Cultures of oligodendrocyte progenitor cells isolated from the trout brain at developmental stage 34 by immunoselection using A2B5-antibodies. **(A,B)** Phase contrast photomicrographs of cultures at 2 DIV (A) and 6 DIV (B). **(C)** A2B5 immunofluorescence labeling of cultured cells at 2 DIV. **(D)** Immunostaining with 6D2-antibodies after 7 DIV. Scale bars indicate 45 μm.

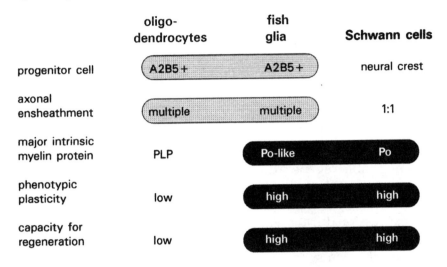

	oligo-dendrocytes	fish glia	Schwann cells
progenitor cell	A2B5 +	A2B5 +	neural crest
axonal ensheathment	multiple	multiple	1:1
major intrinsic myelin protein	PLP	Po-like	Po
phenotypic plasticity	low	high	high
capacity for regeneration	low	high	high

Fig. 7. Comparison of the cellular properties of myelin-forming cells in the trout CNS with those of mammalian oligodendrocytes and Schwann cells.

treatment with established growth factors of the mammalian brain such as basic fibroblast growth factor (bFGF) or platelet-derived growth factor (PDGF). Hence the characterization of growth and differentiation factors acting on fish oligodendrocytes remains an intriguing challenge for the near future.

6. Conclusion

The myelin-forming cells of trout CNS in terms of morphology as well as regarding their cell lineage relationships closely resemble mammalian oligodendrocytes, whereas in their molecular phenotype, including the regulation of myelin gene expression, they exhibit striking similarities with mammalian Schwann cells (Fig. 7):

1. Trout oligodendrocytes do not express proteolipid protein, which is an established oligodendrocyte marker in the mammalian brain (Agrawal et al., 1977), but synthesize instead two P_0-like glycoproteins as major constituents of the myelin sheath;
2. Similar as is known for Schwann cells, immature oligodendrocytes of trout are not able to properly differentiate in vitro in the absence of axonal contacts, but seem to require appropriate inducing signals from other cells, e.g., neurons;

3. Mature oligodendrocytes rapidly cease expression of their myelin-related molecules in dissociated cell culture—a process that is regulated at the level of nucleic acids;
4. In contrast to mammalian oligodendrocytes, but similar with Schwann cells, fish oligodendrocytes are able to remyelinate axons after injury (Wolburg, 1981) and seem to promote regenerative neuritic outgrowth in vitro (Bastmeyer et al., 1991).

Taking these findings together, it is tempting to speculate that the remarkable regeneration potential common to both cell types is connected with their high degree of phenotypic plasticity as revealed in cell culture. Therefore, a detailed analysis of myelin gene expression during regeneration of fish CNS fiber tracts would be of great future interest. Since it can be expected that transcriptional mechanisms play a major part in controlling the activity of myelin genes during development and regeneration studies on the characterization of corresponding cis- and trans-acting regulatory elements thereby will be of crucial importance.

Acknowledgments

This work was supported by the Deutsche Forschungsgemeinschaft (grants Je 115/4-2, SFB 171/C13). The authors would like to thank Annette Schulte and Ursula Mädler for their kind help in preparing the manuscript.

References

Abney ER Bartlett PF Raff MC (1981) Astrocytes, ependymal cells and oligodendrocytes develop on schedule in dissociated cell cultures of embryonic rat brain. Dev Biol 83:301–310

Agrawal HC Hartman BK Shearer WT Kalmbach S Margolis FL (1977) Purification and immunohistochemical localization of rat brain myelin proteolipid. J Neurochem 28:495–508

Bansal R Pfeiffer SE (1989) Reversible inhibition of oligodendrocyte progenitor differentiation by a monoclonal antibody against surface galactolipids. Proc Natl Acad Sci USA 86:6181–6185.

Bastmeyer M Beckmannn M Schwab ME Stuermer CAO (1991) Growth of regenerating goldfish axons is inhibited by rat oligodendrocytes and CNS myelin but not goldfish optic nerve tract oligodendrocyte-like cells and fish CNS myelin. J Neurosci 11:626–640.

Braun PE (1984) Molecular organization of myelin, in Myelin (Morell P, ed.), Plenum, New York, pp. 97–116.

Brockes IP Raff MC Nishiguchi DH Winter J (1979) Studies on cultured rat Schwann cells. III. Assays for peripheral myelin protein. J Neurocytol 9:66–77.

Brophy PJ Boccaccio GL Colman DR (1993) The distribution of myelin basic protein mRNAs within myelinating oligodendrocytes. TINS 12:515–521.

Bunge RP Bunge MB Eldridge CF (1986) Linkage between axonal ensheathment and basal lamina production by Schwann cells. Ann Rev Neurosci 9:305–328.

Carnegie PR Dunkley PR (1975) Basic proteins of central and peripheral nervous system myelin. Adv Neurochem 1:95–135

Colman DR Kreibich G Frey AB Sabatini DD (1982) Synthesis and incorporation of myelin polypeptides into CNS myelin. J Cell Biol 95:598–608.

D'Urso D Brophy PJ Staugaitis SM Gillespie CS Frey AB Stempak JG Colman DR (1990) Protein zero of peripheral nerve myelin: biosynthesis, membrane insertion and evidence for homotypic interaction. Neuron 2:449–460.

deFerra F Engh H Hudson L Kamholz J Puckett C Molineaux S Lazzarini RA (1985) Alternative splicing accounts for the four forms of myelin basic protein. Cell 43:721–727.

Dyer CA Benjamins JA (1990) Glycolipids and transmembrane signalling: antibodies to galactocerebroside cause an influx of calcium in oligodendrocytes. J Cell Biol 111:625–633.

Everly JL Brady RO Quarles RH (1973) Evidence that the major protein in rat sciatic nerve myelin is a glycoprotein. J Neurochem 21:329–334.

Filbin MT Walsh FS Trapp BD Pizzey JA Tennekoon GI (1990) Role of myelin PO protein as a homophilic adhesion molecule. Nature 344:871.

Gard AL Pfeiffer SE (1989) Oligodendrocyte progenitors isolated directly from developing telencephalon at an specific phenotypic stage: myelinogenic potential in a defined environment. Development 106:119–132.

Jeserich G (1981) A morphological and biochemical study of myelinogenesis in fish brain. Dev Neurosci 4:373–381.

Jeserich G (1983) Protein analysis of myelin isolated from the CNS of fish: developmental and species comparisons. Neurochem Res 8:957–969.

Jeserich G Müller A Jacque C (1990a) Developmental expression of myelin proteins by oligodendrocytes in the CNS of trout. Dev Brain Res 51:27–34.

Jeserich G Rauen T (1990) Cell cultures enriched in oligodendrocytes from the central nervous system of trout in terms of phenotypic expression exhibit parallels with cultured Schwann cells. Glia 3:65–74.

Jeserich G Rauen T Stratmann A (1990b) Myelin and myelin-forming cells in the brain of fish—A cell culture approach, in Celluar and Molecular Biology of Myelination (Jeserich G Althaus H Waehneldt TV, eds.), Springer, Heidelberg, pp. 343–359.

Jeserich G Waehneldt TV (1986a) Bony fish myelin: evidence for common major structural glycoproteins in central and peripheral myelin of trout. J Neurochem 46:525–533.

Jeserich G Waehneldt TV (1986b) Characterization of antibodies against major fish CNS myelin proteins: immunoblot analysis and immunohistochemical localization of 36K and IP2 proteins in trout nerve tissue. J Neurosci Res 15:147–158.

Jeserich G Waehneldt TV (1987) Antigenic sites common to major fish myelin glycoproteins (IP) and to major tretrapod PNS myelin glycoprotein Po reside in the amino acid chains. Neurochem Res 12:821–825.

Jeserich G Stratmann A (1992) In vitro differentiation of trout oligodendrocytes: evidence for an A2B5-positive origin. Dev Brain Res 67:27–35.

Jeserich G Strelau J Tönnies R (1993) An immunomagnetic cell separation technique for isolating oligodendroglial progenitor cells of trout CNS. Neuroprotocols: A comp. Meth Neurosci. 2:219–224.

Jessen KR Mirksy R (1991) Schwann cell precursors and their development. Glia 4:185–194.

Kruger L Maxwell DS (1967) Comparative fine structure of vertebrate neuroglia: Teleosts and reptiles. J Comp Neurol 129:115–124.

Laursen RA Samiullah A Lees MB (1984) The structure of bovine brain myelin proteolipid and its organization in myelin. Proc Nat Acad Sci USA 81:2912–2916.

Le Douarin NM (1986) Cell line segregation during peripheral nervous system ontogeny. Science 231:1515–1522.

Lees MB Brostoff SW (1984) Proteins of myelin, in Myelin (Morell P, ed.), Plenum, New York, pp. 197–224.

Lemke G Axel R (1985) Isolation and sequence of a cDNA encoding the major structural protein of peripheral myelin. Cell 40:501–508.

Lemke G Chao MC (1988) Axons regulate Schwann cell expression of the major myelin and NGF receptor genes. Development 102:499–504.

Lemke G Lamar E Patterson J (1988) Isolation and analysis of the gene encoding myelin protein zero. Neuron 1:73–83.

Lemke, G (1988) Unwrapping the genes of myelin. Neuron 1:535–543.

Mirsky E., Winter J Abney ER Pruss RM Gavrilovic J Raff MC (1980) Myelin specific proteins and glycolipids in rat Schwann cells and oligodendrocytes in cultures. J Cell Biol 84:483–494.

Monge M Kadiiski D Jacque C Zalc, B (1986) Oligodendroglial expression and deposition of four major myelin constituents in the myelin sheath during development. An in vivo study. Dev Neurosci 8:222–235.

Nave KA Milner RJ (1989) Proteolipid proteins: structure and genetic expression in normal and myelin-deficient mutant mice. Crit Rev Neurobiol 5:65–69.

Nona SN Shehab SAS Stafford CA Cronly-Dillon JR (1989) Glial fibrillary acid protein (GFAP) from goldfish: its localisation in visual pathway. Glia 2:189–200.

Omlin FX Webster HdeF Palkovits CG Cohen SR (1988) Immunocytochemical localization of basic protein in major dense line regions of central and peripheral myelin. J Cell Biol 95:242–248.

Peters A Palay S Webster HdeF (1976) The Fine Structure of the Nervous System. W. B. Saunders, Philadelphia.

Pfeiffer SE Warrington AE Bansal R (1993) The oligodendrocyte and its many cellular processes. Trends in Cell Biol 6: 191–197.

Privat A Jacque C Bourre JM Dupouey P Baumann N (1987) Absence of the major dense line in myelin of the mutant mouse "shiverer." Neurosci Lett 12:107–112.

Raff MC Miller RH , Noble M (1983) A glial progenitor cell that develops in vitro into an astrocyte or oligodendrocyte depending on culture medium. Nature (Lond.) 303:390–396

Raine CS (1984) Morphology of myelin and myelination, in Myelin (Morell P, ed.), Plenum, New York, pp. 1–50.

Ritchie JM (1984) Physiological basis of conduction in myelinated nerve fibres, in Myelin (Morell P, ed.), Plenum, New York, pp. 117–145.

Roach A Boylan K Horvath S Prusiner SB Hood LE (1983) Characterization of cloned cDNA representing rat myelin basic protein: Absence of expression in brain of shiverer mutant mice. Cell 34:799–806

Roots BI Agrawal D Weir G Agrawal HC (1984) Immunoblot identification of 13.5 kilodalton myelin basic protein in goldfish brain myelin. J Neurochem 43:1421–1424.

Rumsby MG Crang AJ (1977) The myelin sheath: a structural examination, in The Synthesis, Assembly and Turnover of Cell Surface Components (Poste G Nicolson GL, eds.), North-Holland, Amsterdam, pp. 247–362.

Sternberger NH Itoyama Y Kies NW Webster H deF (1978) Immunocytochemical method to identify basic protein in myelin-forming oligodendrocytes of newborn rat CNS. J Neurocytol 7:251–263.

Stoffel W Hillen H Giersiefen R (1984) Structure and molecular arrangement of proteolipid protein of central nervous system myelin. Proc Natl Acad Sci USA 81:5012–5016.

Stratmann A Jeserich G (1995) Molecular cloning and tissue expression of a cDNA encoding IP1-a P_0-like glycoprotein of trout CNS myelin. J Neurochem, in press.

Trapp BD Hauer P Lemke G (1988) Axonal regulation of myelin protein mRNA levels in actively myelinating Schwann cells. J Neurosci 8:3515–3521.

Vernier JM (1969) Table chronologique du développement embryonaire de la truite arc-en-ciel Salmo gairdneri, Rich. 1836, Ann Embryol Morphog 2:495–520.

Waehneldt TV Jeserich G (1984) Biochemical characterization of the central nervous system myelin proteins of the rainbow trout, Salmo gairdneri. Brain Res 309:127–134.

Waehneldt TV Matthieu JM Jeserich G (1986a) Appearance of myelin proteins during vertebrate evolution. Neurochem Int 9:463–474.

Waehneldt TV Stocklas S Jeserich G Matthieu JM (1986b) Central nervous system myelin of teleosts: comparative electrophoretic analysis of its proteins by staining and immunoblotting. Comp Biochem Physiol 84B:273–278.

Wolburg H (1981) Myelination and remyelination in the regenerating visual system of the goldfish. Exp Brain Res 43:199–206.

Wood P Bunge RP (1984) The biology of the oligodendrocyte, in Advances in Neurochemistry, vol. 5: Oligodendroglia (Norton WT, ed.), Plenum, New York, pp. 1–46.

Zeller NK Behar TN Dubois-Dalcq ME Lazzarini RA (1985) The timely expression of myelin basic protein gene in cultured rat brain oligodendrocytes is independent of continuous neuronal influences. J Neurosci 5:2955–2962.

Index

DATE DUE